THE END OF THE WORLD AND THE LAST GOD

The END OF THE WORLD AND THE LAST GOD

PIERRE-HENRI D'ARGENSON

TRANSLATED BY

JAMES CHRISTIE

FIRST HILL BOOKS
An imprint of Wimbledon Publishing Company

This edition first published in UK and USA 2021
by FIRST HILL BOOKS
75–76 Blackfriars Road, London SE1 8HA, UK
or PO Box 9779, London SW19 7ZG, UK
and
244 Madison Ave #116, New York, NY 10016, USA

First published in French as *La fin du monde et le dernier dieu, un nouvel horizon pour l'humanité*
© Editions Liber, Montréal, 2018

Translated into English by Mr. James Christie (james.christie.cit@gmail.com)

British Library Cataloguing-in-Publication Data
A catalogue record for this book is available from the British Library.

Library of Congress Control Number: 2021946518

ISBN-13: 978-1-83998-187-6 (Hbk)
ISBN-10: 1-83998-187-3 (Hbk)

This title is also available as an e-book.

Many may still deem this idea as pure folly, but I remain convinced that the most important discovery to be made, one that may well be reached during this century, is the invention of flight. Through this, man will travel faster and more comfortably, even transporting goods aboard vast flying vessels. There will be armed air forces. Our current fortifications will be rendered obsolete […]. However, our artillerymen will of course learn to hit flying targets. Our kingdom will therefore need a new Secretary of State in charge of the air force.

Marquis d'Argenson, *Mémoires* (written between 1728 and 1757)[1]

We've always defined ourselves by the ability to overcome the impossible. And we count these moments. These moments when we dared to aim higher, to break barriers, to reach for the stars, to make the unknown known. We count these moments as our proudest achievements. But we lost all that. Or perhaps we've just forgotten that we are still pioneers. And we've barely begun. And that our greatest accomplishments cannot be behind us, because our destiny lies above us.

Joseph Cooper (Matthew McConaughey), *Interstellar* (2014, Christopher Nolan)

1 This quote can be found engraved up in French on Aviation Wall in Felicity, California, which charts the key landmarks of French aeronautics. The first manned flight took place in a hot air balloon made by the Montgolfier brothers in Paris on November 21, 1783. French military aviation arrived in 1909, and France's Air Ministry was created in 1928.

CONTENTS

PROLOGUE

From our first steps through the African savannah 2 million years ago, right up until we started to make our way into the stratosphere, we humans never truly envisaged that our destiny could lie elsewhere than on Earth. The idea that we could one day be meant to leave the planet where we were born was just not conceivable by most people in the past. Their imagination was peopled by angels, demons or spirits, not by aliens and space battles. For them, the celestial heavens remained a magnificent backdrop thought up by God or competing divinities, and absolutely nothing in any antique cosmogony suggests that a possible departure from Earth could one day be humanity's greatest adventure. Medieval man would have had a spiritual breakdown if he had been invited to step aboard a spaceship and take a voyage to colonize another star system.

Yet things have changed with space conquest. As we gradually discover the existence of exoplanets suited to our wildest dreams, the possibility of a departure from Earth sounds increasingly familiar and exciting. Indeed, as inaccessible as this may seem, at least for the decades to come, the likelihood of interstellar travel should no longer be viewed, given the amazing technological progress we are witnessing every single day, as an unrealistic dream. After years of relative indifference following the exploits of the Apollo missions, space exploration has obviously caught the public's imagination once again, and we're shyly beginning to admit that traveling to the most lavish and furthest-flung reaches of the universe could one day come true.

However, this enthusiasm may well hide a disturbing existential question, with potentially dramatic consequences: what if humankind is in fact bored with life on Earth?

If someone had presented me with that question during a conversation, I would have been slightly troubled, but this only on an intellectual level. But if this awareness of a boring world had come to me through the experience of everyday trials and tribulations, then I would have devoted my full attention to it. This, of course, is exactly what happened. A few years ago, I was depressed at work, but I could not distinguish between a simple "bore-out" or disillusionment concerning my professional life, which can affect anyone, and a more tenacious *ennui*, revealed in a disappointing everyday life but reaching to the very essence of my entire human existence. By slowly tracing the original source of that *ennui*, I discovered that the reason for it was not to be found in anything serious related to my life conditions and perspectives, but in a both unexpected and fateful event that is happening right now: the end of our world.

Upon hearing such words, one is more likely to think back to the death of humanity in one apocalyptic shower of fireballs, with yawning chasms appearing in the earth and skies rent asunder amidst a Dantesque inferno. But the end of the world I am talking about is already taking place, in a way that is rather less spectacular but just as real, which I will for the moment describe thus: man's potential on Earth has been exhausted. The world has ended, for it has nothing new to offer us anymore. We have discovered every piece of land, tried all sorts of political regime, committed ourselves to all possible kinds of religious beliefs and possibly achieved all the forms of art. We have become old and weary. Indeed, this finite world seems no longer suited to our biology or to our psychology, which were designed by and for exploration. This exhaustion of terrestrial possibilities was always going to happen, for it is in our human nature not only to conquer our surrounding environment but also to stretch, knead and shape our situation to provide us with every possible future, for better or worse. It is humankind that brought on the end of the world, and unless early man had stayed hidden in his cave, this was the inevitable outcome.

Yet if we accept that the thirst for conquest and exploration is at the heart of human nature, we will have to lucidly confront the resulting question: can humanity survive the finitude of our world? Will we hold out long in this cloistered, explored, domesticated, mapped, charted, limited, monitored, ransacked and stripped world—this Earth that has become so transparent and devoid of all mystery?

Up until now, the haunting prospect of a finite world had been nothing more than just another concept. Little by little, I began to realize that this moment is indeed upon us, despite all efforts made to ignore or deny it, resulting in an unvoiced fear and the following nagging doubt: if everything has been tested and explored, what on earth are we going to do now? Upon which goal will we focus our thirst for conquest, our need for the novel, our desire to learn and experiment? Will we head off to conquer space? And if we can't, will we recreate the clean slate we so yearn for by destroying our world and start over anew, as after the Great Flood? Or will we die of boredom, when Earth will have become like a tiger's cage, the biggest open-air zoo in the universe? Knowing that the world can offer us no more is not merely a piece of information; it is a shattering reality to which our bodies and minds will react wildly and, in my view, the biggest challenge humanity will have to face in the near future.

Part of the solution may lie above us, but leaving Earth is easier said than done. The day when our technology can catapult us in one piece to a planet orbiting another star is considerably far off. Just getting to Mars is already a colossal endeavor. Space is not a friendly environment for humans, nor, apparently, for any living being. In 1950, physicist Enrico Fermi came up with a now-famous paradox, which can be summarized as follows: given that the majority of the universe is much older than Earth, extraterrestrial civilizations, if they exist or have existed, should either have visited us by now or have left some visible trace in space. Why then have we not spotted any little green men through our telescopes? Where are the Others? According to American economist Robin Hanson, one of the most recent thinkers to study this question seriously, this can mean only one of two things: either we truly are the only evolved life form in the universe, or assuming that they, too, have evolved from a microbial form to a fully developed life form, all other extraterrestrial civilizations have encountered some mysterious obstacle to interstellar expansion, known as the Great Filter.[2] The question Hanson asks for planet Earth is whether this obstacle lies in our past or in our future. If it lies in our past, we have already overcome this obstacle, and it may well be our destiny to colonize the galaxy. If the barrier lies ahead of us, then it would limit our chances of leaving our own star system, for there is no known proof across the universe that anyone so far has been able to do so.

One could evade this issue simply by saying that we still have plenty of time to develop the technology required for our Great Exodus. However, if we wish to survive as a species, I fear that we don't have much time. Earth's history has taught us that huge cataclysmic exterminations occur with a certain recurrence, resulting in the extinction of species on a massive scale. Earth has experienced five of these in the past. The sixth, brought on by man, is apparently already underway.[3] According to professor Stephen Nelson, a meteorite with a diameter of 1 km strikes Earth once every million years on average, and a meteorite with a 10 km diameter once every 100 million years, causing a catastrophe with the

2 "The Great Filter—Are We Almost Past It?" September 15, 1998, Mason University, https://mason.gmu.edu/~rhanson/greatfilter.html.

3 On this subject, see Elizabeth Kolbert's *The Sixth Extinction: An Unnatural History* (New York: Henry Holt, 2014).

potential to wipe out life on Earth.[4] So, it is statistically likely that humankind will have to face mortal dangers well before being charred to a crisp by solar inflation. Aware of the risks of Armageddon, Europe and the United States have launched missions to identify large asteroids known as near-Earth objects, which could potentially collide with Earth.

One could also consider that it matters little whether in the end we are capable of colonizing space or not, as long as we have our humdrum lives to attend to here on Earth until our planet dies a quiet natural death. But I believe our human nature will not let this happen. We are wanderers, explorers, pioneers, in all fields of life, and we vitally need to expand. I am indeed putting forward the hypothesis that any civilization that fails to escape its planet after having fully explored its geographical and cultural potential will be condemned to an existential death that will soon result in physical death. In other words, if humankind, fully isolated and faced with the perspective of a future limited by the exhaustion of terrestrial possibilities and therefore stripped of all interest, fails to conquer space, we will surely destroy ourselves. Either through war, depression or suicidal destruction, but well before we fall foul of any cataclysm, meteorite, depletion of natural resources, global warming or worldwide epidemic.

The finitude of our world eventually raises an unprecedented issue, especially in the Christian civilization: to oblige humanity to consider the birthplace of Creation, not as a lush and wonderful garden bequeathed by God for the accomplishment of human destiny, but as a prison to flee. This would explain why we are in fact getting so unhappy with our world, up to a point where we may soon find no more reasons to continue to live in it, not only as a species but also as individuals looking for reasons to work hard, hope, love and make children. By shedding light on this pervading worldwide mental condition, I wish to raise the awareness that discovering the nature of the Great Filter, which may not essentially lie in any technological limitation but in the demons and flaws we host in ourselves, might be the only chance to offer humanity a future worth living.

---

4 "Meteorites, Impacts, and Mass Extinction," Tulane University, April 27, 2018, https://www.tulane.edu/~sanelson/Natural_Disasters/impacts.htm.

PART I

THE VERY END OF OUR WORLD

CHAPTER 1

THE END OF GEOGRAPHY AND THE DEMISE OF THE HUMAN BODY

As Chris Impey points out in *Beyond: Our Future in Space*,[1] wanderlust can be considered as one of our most innate emotions. Since humans appeared on Earth, they never stopped exploring, and this is certainly what gave them the best opportunities to be creative and to invent new techniques and technologies. Those things made them the happiest. The history of humankind demonstrates it, but children's dreams and adventures offer the best of confirmations of the extent to which exploring makes us feel alive. But now that there is almost nothing left for us to discover on Earth, a crucial part of our psychological and physical needs finds itself increasingly hard to fulfill. Indeed, our planet's geographical conquest, which accelerated suddenly with the Age of Discovery in the fifteenth century after thousands of years of relatively slow motion, came to an end with the detailed satellite cartography of even the world's darkest nooks and crannies.[2] Only a few regions remain unexplored or barely known: the Mariana Trench, the region of Sierra de Maigualida in Venezuela, the Antarctic, and several mountainous areas of Vietnam, Laos and Indonesia. There are no deserted islands left to visit, no ocean depths left to chart and no mysterious mountains left to climb. Earth has been laid bare and exposed, with each square meter now referenced on Google Earth. There are no more expeditions to explore new lands, so the world is now traveled only by tourists. Geographically, all is henceforth known, including our planet's historical physiognomy, right down to its earliest volcanoes and its first bacterial life forms.

1 Chris Impey, *Beyond: Our Future in Space* (New York: Norton, 2016).

2 One should perhaps speak of an Age of Human Intersection or Unification instead, given that a majority of places discovered after the fifteenth century were, with a few exceptions, not unknown to the human race.

Exploration has always implied a strong physical commitment, along with enormous risks. Throughout humanity's conquest of the world, many people died while trying to cross mountains, rivers, jungles, deserts and seas. But despite our apparent weaknesses (we've got no fur, no shell, no claws or venom), we are actually quite resistant and adaptable, at least enough for our intelligence to compensate for our physical deficiencies. Our bodies have been designed for walking long distances, running and even swimming. There are not many animals as polyvalent as us. So, in addition to frustrating our taste for adventure, the end of the terrestrial exploration era has had an unexpected and desolating consequence on our condition: the obsolescence of our bodies. In a finite world, where there is nothing left to discover, human activity will inevitably lose all tangible form, shifting progressively from the physical world to the virtual, from outdoor activity on the field to mere data processing.

This is something I could experience in my own life. As a child, I had the chance to often go to the countryside and walk and run and fish and muck around as I wanted. The more I grew up, the less I moved, and the more time I spent at home, at school, at university and then at work. One evening, I realized that I was actually spending the better part of every day sat down in a chair, just like every other person who works in an office, and the result was that my body no longer served any purpose. This realization became a source of considerable suffering for me. Upon observing my lifestyle, I suddenly understood that the majority of men and women have become physically useless. In today's postindustrial society, a large number of people are employed in the tertiary sector, doing jobs that require no physical effort whatsoever, including even getting to one's workplace.

In fact, the only real physical effort required for those kinds of jobs consists of sitting still and remaining hunched over one's desk for such a long time that one eventually ends up falling ill. Everything we do in an office is done sitting on our backsides, our bulging eyes staring at screens while we scream silent orders to our fingertips as they tap-dance across our keyboards. If things keep going this way, in a few hundred years' time we will be born shortsighted and hunchbacked, with bulbous heads, 15 fingers on each hand, outsized buttocks and long spindly legs like a spider! Physically speaking, a brain hooked up to electrodes would be able to carry out the same tasks as many urban office workers, whose bodies' only real purpose is to transport their brains. Even this, in a near future, will be rendered useless by artificial intelligence (AI), as humans will be considered as less efficient as robots for many data-processing jobs. Our lives often look like a string of places to park our rear ends, from meeting rooms to subway seats, armchairs to benches, stools to couches, office desks to movie theaters and cars to suburban commuter trains. We sit down on public transport or in our cars as we commute to work, sitting down once again in front of our computers, smartphones, televisions or cinema screens for such long periods of time that we may end up suffering from gluteal amnesia.[3] And then we go to bed. And to prevent our nerves from snapping, we spend time at the gym, busying ourselves like competing hamsters, or go out running, pushing our bodies to the point at which our faces crease up in agony, as if we were being hunted down by a raging *Tyrannosaurus rex*.

Technical progress, our planet's digital conquest and the advent of the information generation all took place at the same speed in time, for they all originated from the very same source. Technology and automation have taken over numerous physical tasks to such an extent that only a minority of people continue to make a living from manual work, and this rarely well-paid. In order for our consumer society to function correctly, we still need people to put in the hard graft: laborers, workers, soldiers, those working behind the scenes when we buy our groceries, those who keep our fridges full and our stomachs fed and enable us to own modern appliances and to live in peace and security. We are speaking of archetypal categories here: I am not forgetting the policemen, firemen, garbage collectors, sewer workers, factory workers, woodcutters, herdsmen, warehouse handlers and all professions for whom the body is still heavily required. They work in the shadows, their vocations downgraded in comparison with professions related to digital finance and online trade. But this remains a trick of economics—society may not be able to operate without them, but it can function just fine with the illusion of their absence and, consequently, their irrelevance. Never in history has the world seen such a huge gap between manual and intellectual tasks, including geographically—engineers and consultants keep well away from the factory floor these days.[4]

---

3 Or "dormant butt syndrome," a poetic term coined by Chris Kolba, a physical therapist at Ohio State Wexner Medical Center, which refers to the weakness of the gluteal muscles caused by prolonged sitting, https://wexnermedical.osu.edu/mediaroom/pressreleaselisting/dormant-butt-syndrome-mmr.

4 Historically, the power of machines progressively replaced the strength of men in many areas, paving the way toward the reign of the tertiary sector. Ironically, AI could soon in turn put an end to the domination of the human intellect, rendering most of us obsolete, with the exception of a tiny high IQ elite. Elon Musk's neuroenhancement project Neuralink may only be able to delay the timetable (Neuralink aims at developing implantable brain–computer interfaces, changing us into cyborgs supposedly capable of competing with AI).

At the turn of the century, politicians decided to sugarcoat this tragedy by giving it a cozy euphemism, the "advent of the knowledge economy," placing a chaste veil over the fact that most modern workers no longer know what to do with their own four limbs. One may argue that this phenomenon is in itself not new and that we are simply witnessing the enlargement of the class of scribes of old. But I would point out that until at least the eighteenth century, kings or queens and their advisers were present on the battlefield, something we no longer see today, and it was commonplace for their own sons to fight and risk their lives. Over the past few decades, have we seen but one defense secretary merely assisting at a battle? Or risking their children's lives in one of the wars they started? Even though they managed to retain an air of a privileged caste, the "technocrats" of old still carried out their roles in physically harsh conditions, for example, when they had to inspect garrisons, camps and strongholds—places often rife with disease. The world's political and financial elite of today may indeed face considerable levels of stress, but their conditions are nothing like those of the previous centuries. The last of this mold were probably de Gaulle, Churchill and Eisenhower, who all had a son fighting during the Second World War. Since them, the power has been squarely in the hands of the dainty and the delicate.

In contrast to this immobile atmosphere, the modern hiker is carrying out, without knowing it, an act of commemoration. In the same way that a hamster is given an exercise wheel to keep fit, humans can't help using their basic biological capacity: to walk, thus doing exactly what our bodies and souls were made for. Heading off on a 10-day trek in the mountains with no other goal but to walk the footpaths with a backpack is something that would have puzzled even our recent ancestors. It's also fascinating that the leisure pursuits of the Western middle classes in the twenty-first century are the things people with less money and modernity do simply to stay alive: gardening, walking, running, fishing, shooting, riding, picking up heavy objects and putting them down again. The explosion of modern tourism is also a symptom of the collective angst that has gripped humanity since the moment our planet had nothing mysterious left to offer. We "get off" on long-haul flights and rack up exotic destinations and extreme sports that bring us the hormonal stimulation that we need to compensate for the disappearance of any real-life adventure.[5]

---

5 The Hollywood movie *The Secret Life of Walter Mitty* (Ben Stiller, 2013) provides a good insight of the frustrated exploration and adventure fantasies of modern office workers.

Over a period of a thousand years, we have gone from a society in which almost everyone had a profession that put their body to the test, exposing and manhandling it, to a world where the (digital) quill has taken over. "Greek and Roman societies did not dissociate biology from politics. The body, the city-state, the sea, the fields, war and works of art were all confronted with a single vitality, all exposed to the same risk of sterility, subject to the same appeals for fertility," according to Pascal Quignard.[6] Our computers have finished off the job started by factory production lines: the great separation. Man has forsaken both nature and his own nature, and the reason he treats them so badly is that he has become foreign to both. But yet he knows that he is slowly dying inside, withering away in meetings, plastic environments with dirty carpets and high-rise air-conditioned office buildings that sprout up in business districts like concrete plants. In these tin cans, everything is shiny, metallic and reflective—it is ironic then that when we are fired, we "get the can." We breathe in exhaust fumes, sweat and deodorant, inhaling dust mites and sidestepping pigeon droppings. We create no art, no exploit is applauded and no memories live on in these places. Everyone within the hierarchy, from the top to the bottom, does the same job: the sorting, compiling, forwarding, analyzing, distributing, commenting, withholding and transforming of information, using concepts that have relevance to their brain alone, but are of no interest to their body.[7]

The imposed inertia of modern labor, which is a consequence of our explored world, has generated, I believe, a specific male malaise. The economy's inevitable shift toward the service sector has indeed diminished the ancestral distinction of man, that one feature which had somehow justified his existence since the dawn of time: his physical strength. During the exploration era, the male sex used to represent survival through the strength and speed of his muscles and biological makeup, but has been rendered obsolete in today's world of stationary typists. Secretarial work was a profession looked down upon by the ruling masculine elite when women massively entered the modern working world in the mid-twentieth century, and yet everyone is now a typist.

We should feel indeed for modern fathers: how can their importance fail to diminish in the eyes of their children when there is nothing left that connects

---

6 Pascal Quignard, *Sex and Terror*, trans. Chris Turner (London: Seagull Books, 2012), 211.

7 For a brilliant manifesto against the educational imperative of turning everyone into a "knowledge worker," see Matthew B. Crawford, *Shop Class as Soulcraft: An Inquiry into the Value of Work* (New York: Penguin Books, 2010), which calls for a reappraisal of the merits of skilled manual labor.

what they do at the office to fairy tales, heroes and folklore? When there are neither quivers nor arrows, neither swords nor shields, neither forge or furnace nor pottery or clay, neither rivers nor huts, neither aircrafts nor ships? When there are neither companions nor comrades, brothers-in-arms, sidekicks or accomplices? There is nothing in common between the trapper, knight, explorer, blacksmith, adventurer and hunter and the dad who works in an office, doing a job his children will probably not fully understand until they are adults. There are no heroes in office life. The human male's deep-set, reptilian and ancient desire was to go out hunting, with hunger gnawing his stomach in the hope of returning to his tribe bearing game, fruit and fish. His reward was recognition from his spouse and children, their pride and happiness. They needed him; he was useful, he was brave. Since the arrival of the supermarket, man has needed to find another role to justify his existence, and so he creates imaginary needs and concepts to replace the victuals of old. He pretends to go out hunting, whereas in reality he is simply tapping away on a computer, alone. He comes home late on purpose, not because he needs to, but to prove to his family that he is returning from afar, that he has made it all the way to the waterfall, the prairie or the foot of the mountain. He doesn't bring anything back with him, but the reflex still remains. But the truth is: man has fulfilled his role of exploring and conquering, and the modern workplace has seriously sapped his motivation. As I wrote earlier, the world wouldn't go round without the many highly physical professions that exist. But in terms of real adventure, the vast majority of men now have to make do with gathering expeditions to the supermarket.

The obsolescence of men's physical usefulness may be more pronounced in Western societies, but this phenomenon can be observed all over the world, crossing both cultures and borders. This is the biggest change to affect our civilization over the past 500 years. I am not saying that in old times women's bodies were not put to the test, but that men's physical advantage, supposed or real, was part of their core identity, and this contributed to shaping long-term sex representations and stereotypes. Between *Homo habilis* and the Middle Ages, it was almost impossible for any man who had not risked his skin, albeit but a little, to reach any social position whatsoever, and this includes several philosophers: Socrates was a hoplite during three military campaigns, Plato was a sportsman and traveler, Avicenna and Averroes were doctors and Descartes joined up to fight in King Maximilian of Bavaria's army. Right up to the Second World War, it was unimaginable to affirm one's virility without having fought for one's country (there were exceptions of course, but they proved the rule). Wars and travels were what made a man. Today, the majority of men have bodies they no longer use, except maybe to push their shopping carts around. In the glass towers of major companies, walking the long corridors of public administrative departments, they look more like well-behaved dressmakers, embroiderers and spinsters, only using their bodies for sports and leisure. Their muscles become weak, their prowess withers away[8] and their fighting spirit dies out. With a bit of exaggeration, we could almost suggest that transporting semen is more or less all that is required of men today, but this only until the day they manage to invent synthetic sperm.[9]

Though this forced immobility of modern work seems to be more badly felt by men,[10] I am convinced that it will soon become as unbearable for women as it is today for men, thanks to the feminization of some symbolic positions in very "physical" jobs that were formerly reserved for men: submarine crews, fighter pilots, astronauts and so on. But the key point remains the following: while the role of women has been extended far beyond that of nurturing children and managing the home, recent times have seen an entrenchment of men's historical raison d'être, and this is closely linked to the finalization of our planet's conquest. What is left to do on Earth looks much more like interior design than shell building, and I don't think men and women will be forever happy to change the color of the wallpaper or the place of the sofa.

This leads me to the key question that nobody dares to confront: how can we ever survive in such an antilife? Where is the adventure in walking down a corridor? What close bonds can be forged during a meeting? Where are the oceans, lands to discover, forests to cross, wild animals to face and caves to explore? In a world where our bodies are no longer needed, can we truly

---

8 Quite literally: according to an Italian study conducted by Professor Carlo Foresta (University Hospital of Padoue), the average penis length had gone from 9.7 cm in 1948 to 8.9 cm in 2012. But this information has been reported only from Italian newspapers, and no study papers have been published. Should this phenomenon be confirmed by future research, it will not ease the masculinity crisis.

9 I do not wish to scare male readers too much, but this is in fact already the case: on September 17, 2015, a biotech company in Lyon, France, called Kallisten announced they had created complete human spermatozoids in vitro. And in Denmark, 1 in 10 babies born from a sperm donor are raised by a single mother who has chosen to start a family without a father.

10 This can be seen in the "angry white male" unrest that became visible when the industrial sector collapsed in some areas of the United States, notably the Rust Belt. See, for instance, Michael Kimmel, *Angry White Men: American Masculinity at the End of an Era* (New York: Nation Books, 2013).

build human relationships in situations based uniquely on the transmission of information?

During most of the so-called interaction with our fellow men, all we have left to exchange is data—even trading goods no longer requires any in-depth dialogue, as it invariably goes through a middleman: we no longer buy something from the person who made it. People can work together in the same company for years without ever making a true connection, for they never fully put themselves in the hands of the other. Real teamwork is only ever observed when actions are connected to dangerous or grueling activities such as waging war, hunting, fishing, exploring, rescuing, conducting operations, engineering, animal-rearing, farming and so forth. The strongest human bonds, those which bring meaning to existence, are created when we are physically acting together, for this is the only occasion when we really depend on others and we count for others. It is therefore unsurprising that man is depressed by a world in which it is increasingly difficult to attain his greatest sense of fulfillment: to feel genuinely connected to others.[11]

Yet if our bodies no longer serve any purpose, what will we do with them? "But this is not so," you may venture, "the body has never been so idolized!" Yes, the body is indeed idolized, but because we are blind to the way it has been made redundant in the real world. The body is today little more than a statue to be waxed, shined and polished, and displayed in a shop window all day long before being hidden away at night. As the body's real usage wanes, value is placed on the cosmetic, the aesthetic and the make-believe, for the human's real body has been stripped of its vitality. Coincidentally, the end of conscription led to the disappearance of all forms of physical differentiation between the generations.[12] Thirty-somethings may spend time at the gym, but fundamentally they live their lives like older people. And yet, under our surface of *sapiens* coating, we remain *Homo habilis* with strong *erectus* tendencies. For the first time in our evolution, this finite world is no longer suited to our biology. Having spent tens of thousands of years overcoming frostbite, volcanic eruptions, jungle diseases and wild

11 As to the question whether online social networking offers greater friendly connection, anthropologist Robin Dunbar found out that the average number of friends on Facebook approximates the natural size of personal social networks—about 150 individuals (among whom only five can be considered as the best friends, including family members), thus reflecting the fact that real relationships require at least occasional face-to-face interaction to maintain them, http://rsos.royalsocietypublishing.org/content/3/1/150292.

12 For instance, 1962 in the United Kingdom, 1973 in the United States, and 2002 in France.

animals, the human body has now exhausted its role and has become useless. It is not only the body that has given up: in a worn-out world with neither appeal nor options, our minds are bound to go round in circles, driven mad by analyzing our own analyses. What heart or soul would wish to live long in such a suffocating environment? We need to find a way out.

CHAPTER 2

THE DECLINE OF EROS

In addition to creating specific masculine disarray, the inevitable decline of men's physical utility in a tertiary and digital society is actually erasing one of our species' most immemorial traits: the difference between the sexes. As it is accompanied by the rise in power for women in domains formerly exclusive to men,[1] it has become increasingly difficult to distinguish between men and women, right down to the very clothes they wear. Yet if most men no longer have any physical role to play in our explored and domesticated world, can they still truly remain men? Can women still hold any desire for the sedate office workers that so many men have become? Can our sexual alterity biologically survive this physical and mental interchangeability between men and women?

Those questions may sound controversial, for one could instantly contend, firstly, that they are based on the assumption that desire can arise only from

---

1 For the first time in history, the weaker sex may well indeed become the stronger. This is the thesis notably defended in two books, *The End of Men* by Hanna Rosin (New York: Riverhead Hardcover, 2012) and *The Richer Sex* by Liza Mundi (New York: Simon and Schuster, 2012), which highlight a certain number of striking changes seen around the world, including the following: girls between the ages of 14 and 18 perform better in IQ tests than boys; the majority of college students are now female, studying prestigious academic courses with better results than men; women are more qualified than men and attach a greater importance to high-paid careers than men; women are increasingly the family's main breadwinner; and in developed countries, there are now more women marrying men with lower levels of education than vice versa. In the United States, these developments are even greater pronounced: private universities now have to carry out discreet positive discrimination in favor of men, for boys tend to fall behind more at school (also the case in the United Kingdom); women under the age of 30 earn higher salaries on average than men; the median income for working-class American women is higher than that of men from the same background; and the majority of those affected by unemployment are men. This trend, which can also be observed in developing countries, represents one of the greatest upheavals that humankind has ever seen.

gender differentiation, and secondly, that we should in fact be delighted with the growing similarity between men and women, as a decisive step toward closing the gender gap. In a premonitory essay, French feminist Élisabeth Badinter was already rejoicing in the convergence of the sexes, convinced that it heralded the end of the war between the sexes and the birth of a new transgender humanity:

> But the equality, which is now coming about engenders a resemblance that puts an end to this war. Now that the protagonists like to think of themselves as the 'whole' of humanity, each side is in a better position to understand the Other, which has become its shadow. The feelings uniting this couple of mutants can only change their nature. Strangeness disappears and becomes 'familiarity'. We may perhaps lose something of our passion and desire thereby, but we shall gain the sort of tenderness and co-operation that can unite the members of a single family: the mother and her child, the brother and sister. [...] In short, all those who have laid down their arms.[2]

She also writes:

> Our mutant hearts no longer seek the torments of desire. We could almost say they wouldn't know what to do with them. The resemblance model goes together with the eradication of desire. [...] The lovers are bothers. The sexual relationship becomes one of the components of this fraternal relationship which contains something like a slight of incest.[3]

Badinter therefore admits that desire mainly proceeds from estrangement and that we cannot have both gender resemblance and desire for the opposite sex. Although she sees the final trade-off as a positive thing, one question remains: if all desire dies out, will humanity have any reason to live on?

In any event, never in history have human relations been so desexualized, and I believe that this is a direct consequence of our world's exhaustion. According to a study published in the *Archives of Sexual Behavior*, Americans who were married or living together had sex 16 fewer times per year in 2010–14 compared to 2000–2004.[4] Overall, Americans had sex about nine fewer times

---

2 Élisabeth Badinter, *The Unopposite Sex*, trans. Barbara Wright (New York: HarperCollins, 1989), 150–51.

3 Ibid., 206–8.

4 J. M. Twenge, R. A. Sherman and B. E. Wells, "Declines in Sexual Frequency among American Adults, 1989–2014," *Archives of Sexual Behavior* 46, no. 8 (March 6, 2017): 2389–401.

per year in 2010–14 compared to 1995–99. Similar trends can be found in other countries. These figures are by themselves very telling, but it is in the most banal of everyday occurrences that I have spotted signs, nay proof, of Eros's demise on Earth.

While distractedly flicking through waiting-room magazines, anyone can see that humanity is putting a great deal of effort into carrying out endless surveys, studies and polls to track even the slightest tiniest beginnings of a hint of a drop in sexual activity, be it a "loss of desire" in women, a "waning libido" for men, the boom in asexuality or abstinence in developed countries—Japan appearing to be an extreme example of this.[5] No secret governmental organizations have been set up to fight this, and yet an army of militants seem to have declared war on this global epidemic of hypoactive sexual desire, using all aspects of human psychology as leverage: reprimand ("go and see a marriage guidance counsellor"), blackmail ("if you don't make love often enough, then you're not normal"), manipulation ("it's great for morale!"), medication ("take Viagra!"), transference ("buy a sex toy") or an invitation to return to the adventurous era we know, deep down, we've left behind ("why not tie him up and blindfold him?").[6] Where sex is concerned, the one commandment that resonates across the planet is clear: never ever let him refuse. When observing the frequency at which the subject is broached, it would appear that the waning of our sexual desire is humanity's number one preoccupation, way ahead of global warming, volcanic eruptions, tsunamis, plane crashes, economic crises, wars, epidemics, earthquakes and asteroids. But why can we not just leave sex well alone? And why should the percentage of people who are bored with their sex lives concern us so much?[7] I guess it is because we are touching upon a sensitive subject for the future of humanity, and our pretended offhandedness in the face of this matches the very angst that it causes.

Humans are of course continuing to reproduce, of this there is no doubt. But something has changed. Sex is the same, yet it is somehow different. As Henry Miller said, "The way people talk, the way they walk, the way they dress, the way they eat and where, the way they look at one another, every detail, every gesture

---

5 A Japanese gaming company has even specialized in romance games for women. They can notably experience a virtual 3D marriage.

6 Shown none more so by the success of the famous erotic-masochistic novel by Erika Leonard James, *50 Shades of Grey* (New York: Vintage Books, 2011).

7 In 2006–7, the condom brand Durex conducted a survey of around 26,000 people over the age of 16 across 26 countries, concluding that only 44 percent felt sexually fulfilled. Prehistoric man would have been quite happy with a score like that.

they make reveals the presence or the absence of sex."[8] Today it seems that sex is everywhere, but paradoxically it has disappeared. It is ever-present, but yet has forsaken us. In fact, the all-pervading way sex is represented indirectly reveals its very demise. In economics, a moment comes when the creation of money by central banks no longer has any effect on growth: this is known as the liquidity trap. Similarly, we have fallen into a desire trap. Anaesthetized by the end of the world, we are bombarded with images and sensations of a lewd nature in order to keep us constantly alert, much in the same way adrenalin is given to patients in shock. But those images have been made so banal that they are incapable of stirring true desire, one which grips our insides and gives a voracious hunger for both love and life.

Creative eroticism's final bow came in the world of music during the 1980s, the one remaining arena in Western society where visual transgression, extravagant clothing and flamboyant body language were accepted. The outrageous costumes worn by singers at this time and the ostentatious theatrics that accompanied them made it the carnival to end all carnivals, a decadent and jubilant swansong, before all kinds of lawmakers, industrialists and bureaucrats came in to stifle our world with their leaden shroud of processes, preambles, judgments, subprimes, equity, growth curves and employment rates, bottling our desire and selling it like some cheap mineral water. Screens soon followed, to finish off the job of distracting us from the real world, ending an age when men and women would catch each other's eye, their gazes now fixed on their smartphones, with souls drained and worn out by office life, subway commutes, traffic jams and the haunting specters of unemployment, loneliness, their children's education and work-related stress.

Anyone who is receptive to the conditions of modern-day life and whose five senses are tuned into the beating heart of the world will realize that many things in public life and human relations that may cause sensual friction have been insidiously banned. In nightclubs, we will still dance the night away, but without the carefree abandon for which our bodies yearn. Except in regimented clubs where dance steps are practiced like drills, no one ever dances the waltz, the tango or even the slow dances we used to enjoy during teenage parties. The combination of deafening music, relentless tempo, alcohol and sweltering heat drowns out any sensory appreciation, something of which in reality we are incapable anyway. And, above all, we do not wish to have this modern affliction pointed out to us. We are either too scared or too embarrassed to admit it. All

---

8 Henry Miller, *The World of Sex* (New York: Gove, [1940] 1978), 144.

the beauties of courting have been effaced, and no longer do we breathe in our partner's aroma, explore with tender touches and hold them close as we dance. The courtship ritual requires that we truly live, love, embrace and confront the difference between the sexes, which, despite our efforts, we are able to keep alive only through either pretense or meaningless bestiality.

I am fully aware that people may disagree with this position and argue that human affection is doing just fine, thank you. My aim here is none other than to find words to describe the winds of change I feel blowing across the airs of time, telling me simply that our society, like Earth, is going round and round in circles. As we have nothing new left to achieve, our desire dies out, despite trying in vain to reawaken it with an abundance of glossy advertising and media hype. Our desire needs large open spaces to run free and is partly connected to our instinctive need for conquest, feeding on the fear of the unknown. Nothing stifles desire more than comfort and security. This has been with us since the dawn of time, when our survival depended on our passion for life, perpetuating our kin in a hostile world where death could come at any moment. In primitive times, the sexual act alone could lead us to drop our guard, making us vulnerable to potential dangers. This is what happens at the beginning of *Quest for Fire*,[9] when the two prehistoric protagonists are otherwise engaged by the edge of the swamp and fail to see the advancing hordes until it is too late. From this era when our lust for life and survival instinct were first forged, a lasting and deeply engrained imprint has been left on our psyches: to desire another, one must desire more than just the other. Through our partner, we must desire life itself, the open sea, the forests, the rivers, vertiginous mountaintops, yawning chasms and the starry sky. Wild animals may be protected in a zoo, but rarely reproduce in captivity. Just like them, we need to be exposed to danger to feel this lust and desire. Sexual activity soars during wartime, and herein lies the tragedy of modern utopia: we would ideally like pleasure and desire to flow forth from transparency, tranquility and restraint, but deep down this is simply impossible. The more we control something, the deeper we sink into individual and collective sadness. But how can this be any other way when the end of the world strips away the erotic from our survival instinct?

In *The Tears of Eros*,[10] Georges Bataille enlightens us with the following observation: "The meaning of eroticism escapes anyone who cannot see its

---

9 Directed by Jean-Jacques Annaud (1981).

10 Georges Bataille, *The Tears of Eros*, trans. Peter Connor (San Francisco: City Lights, [1961] 2001), 70.

religious meaning!" Henry Miller followed suit: "To me it seems that sex was best understood, best expressed, in the pagan world, in the world of the primitives and in the religious world."[11] Following this counterintuitive argument that religious fervor and eroticism go together, dechristianization in the Western world could be better explained by the drying up of Western eroticism than by the waning of religious belief. The link between Eros and religion has long been maintained in Christianity itself, as shown by the indecent modillions on Roman churches or the Irish Sheela Na Gigs, representing all kinds of exhibitionist, scatological and pornographic scenes, the erotic ebullience of which has since become unfathomable for us. These whimsical old sculptures reveal to what extent the body has since been ousted from the sacred domain. With this in mind, one should not view the contemporary global obsession with pornography as a sign of debauchery, but as a desperate attempt to recover the sacred Eros of the flesh, devitalized and degraded by a modern and mechanical mercantile world in which religion has succumbed to the same fate, downgrading the physical to something viewed as flawed, weak and sinful, adorable as an icon but shameful in essence. Bataille tells us that the link connecting religion to the erotic is gone, reducing religion to a merely useful morality. No wonder people no longer go to church.

The evolution of arts also accounts for this lassitude of the senses. In a fascinating novel, *The Art of Murder*,[12] Jose Carlos Somoza imagines that in a near future a new form of art will have surged, called hyperdramatic, which consists in real people being used as canvases, painted by great masters down to their most intimate parts, and then posed, rented, sold or even treated as human furniture. Those pieces of live art reflect the ultimate fantasy of a society that has no other desire than to feast on its own flesh. This mixture of narcissistic obsession, commercial "commodification" of human beings and symbolic cannibalism mirrors our time, that of a humanity locked up in a cage and condemned to devour itself. Somoza estimates, in a final note, that in art, everything has been done. Perhaps he was unaware that workshops of human furniture existed already, with strong sexual connotation and designated by the delicate neologism of "forniphilia."[13] But when man becomes art, isn't this the sign that nothing extraneous to him touches him anymore? That he has given up on modeling the world? The history of art is a long-term move from

11 Miller, *World of Sex*, 124.

12 Jose Carlos Somoza, *The Art of* Murder, trans. Nick Caistor (London: Abacus, 2005).

13 Coined by British artist Jeff Gord from the combination of the words "furniture" and "fornication."

figuration to disfiguration, that is, from realistic faces and landscapes to blur, stylization, abstraction, bidet, body art and finally human furniture. Why not then anthropophagy, as a desperate refuge for an agonizing human sensuality? Which radical novelty could emerge in painting or sculpture which could still surprise and thrill our senses? Only architecture seems to offer potential innovative forms (floating buildings, underwater habitats, vegetation walls, space hotels, etc.), whereas contemporary music and literature, despite brief incursions into experimental music and Nouveau Roman, do not seem able to offer alternative ways to the works of old. The human race and its culture do not hold the hotheadedness of passionate young lovers; they are an old couple who savor the classic sounds.

Many would like the question of the intensity of human desire to remain something anecdotal or peripheral to the human condition, nothing more than fodder for magazine sex columnists. This is why it has been placed under a heading of light-hearted subjects, thus making it easier to tackle than if we had to go straight to a section called "the future of humanity." A civilization that loses its eroticism is one from which all life drains away, a sign that the end of the world is here and that we know it full well. The erotic is at the very heart of our human vitality, bound in part to our existential and biological survival. What is eroticism's message to us? It speaks to our body and soul, saying: "this is why you exist." And we do not exist simply for that which is to be revealed to us, but the process of this revelation itself, without which no desire can be stirred. We poor contemporary humans naïvely believe that the object of our desire is hidden behind clothes, when in fact it is the act of ripping the clothing off that really lights our fires! This could explain the hidden meaning to Courbet's famous painting, *The Origin of the World*, an 1866 work that marks the symbolic start of modern pornography: everything that once was secret is now known, and the most striking image to express the complete stripping down of the human condition had to be one of a woman's genitalia, the symbol par excellence of all that is hidden. In the origin of the world, in both its original meaning and that of the eponymous painting, there is, like in today's world, no mystery left to be revealed. The comprehensive exposure of our world and our nudity has abolished all eroticism, and with it the very sexuation upon which our species has depended for millennia.

In a finite and explored world, Eros is bound to collapse. But can we arouse desire and excitement again with something else than pure sex differentiation? We are not comfortable with this question, for our postmodern ideal pretends that we can have it all: desire and resemblance, arousal and security, rapture and predictability. However, I don't think that artificially reviving the gender divide could in any way slow the decline of human desire. At the same time, humans cannot be happy with a post-erotic, post-physical world, in which the only activity for everyone would consist in mere Earth management. Engaging men and women in a great common adventure therefore appears to me as the only way to rekindle the fire of desire.

CHAPTER 3

THE REAL END OF HISTORY

The end of the exploration era has thus deeply thwarted two basic human emotions that are strongly intertwined: wanderlust and Eros. There is another deep aspiration that our finished world could eventually frustrate: our appetite for politics. As Aristotle stated, man is a political animal, and this can also be considered as one of our most ancient characteristics. Aristotle meant not only that man needed to live in a social environment (the *polis*) but also that he could develop his full potential and reach happiness only when participating in the collective making of his destiny throughout history. Yet our political nature could be soon deprived of any possibility of expressing itself, for we have reached not only the end of geography but also the end of history. This concept has been so worn out through philosophical debate, often in an aim to disqualify it, that one no longer pays any attention. The true reason, I believe, is that we are actually disturbed by the thought, so we prefer to dismiss the idea, waving it away with a disdainful hand as if it were a stupid remark from a child. But as the saying goes, the truth comes from the mouth of children.

Unlike with traditional messianism, which situates all major achievements in a future that is either unspecified or constantly postponed, Hegel was the first philosopher to imagine the end of history as a tangible event within our reach, as opposed to a far-off concept. For him, the end of history will arise when the Spirit is revealed within a rational governance in which men live freely and equally, an ideal that he saw beginning to seed in the Germany and France of his era. As such, Hegel's thesis was the very first intuition that the possible combinations of governmental types were limited not only in theory (as has been put forward since Aristotle) but also in practice, as there will be a point in history at which everything will have been tested.

The trials and tribulations of the nineteenth and twentieth centuries showed, provisionally, that Hegel was wrong. But in summer 1989, on the cusp of the fall of communism, political scientist Francis Fukuyama, upon becoming aware of the spread of liberal democracy across the world, revived Hegelian thought by adding a dash of the laws of thermodynamic entropy: just as hot water always ends up cooling, liberal democracy had for him become the inevitable spiritual horizon for political regimes, a thesis that he defended in an article entitled, using a falsely interrogative tone, "The End of History?"[1] The Berlin Wall fell on November 9 of that year, lending his essay a premonitory air and adding to its notoriety. Controversial reactions were swift: in the same way that Marx furiously reacted to Hegel by accusing him of confusing the end of history with the advent of the bourgeois state, Samuel Huntington, Fukuyama's professor, retorted to his pupil in 1993 that history was indeed continuing in the form of a "Clash of Civilizations"[2] resulting in a historical dialectic not unlike that of Marx.[3] Following the medieval aggression between principalities, nineteenth-century conflicts between sovereign states and the ideological wars of the twentieth century, Huntington believed that the twenty-first century would see a cycle of standoffs between the West and other civilizations, the latter rejecting the West's claim that liberal democracy is the pinnacle of global political history. The boiling waters of political, ethnic and religious conflicts could bubble over interminably, Huntington maintained, with no reason to cool down and finish with a Western regime at all.

My position is thus: Fukuyama was right. The end of history has indeed taken place, but not the end that he had imagined. The end of history is not the moment when liberal democracy—or any other ethos—becomes the political regime toward which all countries around the world would lean. The end of history comes with the realization that history has exhausted all forms of political governance. Whatever political regime is adopted by a state or a group of states, it must necessarily be connected to something that has already been tried. In the same way that we have comprehensively indexed all land on Earth, we have already explored all humanly possible forms of political regimes, including the very worst. In light of this, the apparent opposition between

1 Francis Fukuyama, "The End of History?," *National Interest* 16 (Summer 1989), 3–18.

2 Samuel Huntington, "Clash of Civilizations," *Foreign Affairs* 72, no. 3 (Summer 1993), 22–49.

3 Marxist dialectics explain history as a series of social struggles: slaves against freemen, plebeians against patricians, serfs against lords and the proletariat against the bourgeoisie.

Fukuyama and Huntington ceases to exist, for this clash of civilizations, as an obvious regression to ancient forms of political violence, heralds the beginning of a cycle of historical repetition.[4] The clash of civilizations, which is a palpable geopolitical reality and a major step backward for humanity, is therefore proof that we have reached the end of history, and thus, the master and pupil may be reconciled once again.

The term "liberal democracy" illustrates just how traditional historical-political dialectics have been exhausted, for they used to address the struggle between pure political models, those uncontaminated by elements hitherto unknown to political history. Why not just be happy with the term "democracy"? Why feel the need to add the adjective "liberal"? Simply in order to slip in the idea that democracy alone would not be seen in such an appealing light if it were not, first and foremost, liberal, and thus primarily ruled by market forces. In other words, democracy has become acceptable in the eyes of Fukuyama (reflecting here the global intelligentsia's mainstream thought) only if the masses choose to conform to the liberal socioeconomic order, the essence of which is to make the laws of supply and demand prevail over any political order. But why has liberalism become the world's ruling ideology? My hypothesis is that liberalism's apparent (but deceptive) capacity to regulate all aspects of human relations, activities and conflicts without using the traditional mechanisms of political history makes it precisely the best candidate for appeasing our existential angst before the end of history.

I will not engage here in an in-depth discussion about what liberalism really means, for it would lead us to endless philosophical controversy. So, for the purpose of my reflection here, I will use Jean-Claude Michéa's critical approach to liberalism as it is put in practice today,[5] namely, a "radical transformation of the human order," relying on three underlying premises. The first states that society can achieve peace only when people's sole remaining concern is their own material well-being. That is, when humanity has abandoned all heartfelt notions of all matters of historical importance: political ideologies, religions, culture, values—in short, all immaterial affiliations around which conflicts have centered throughout history. The second premise upholds that the best way to

---

4 I have to admit that this linear view of history may seem true and worrying only from a Western perspective. Many Eastern mythologies, seeing history as a never-ending wheel, would even consider the concept of history as nonsensical as a flat-Earth theory.

5 See specifically Jean-Claude Michéa, *The Realm of Lesser Evil*, trans. David Fernbach (Cambridge: Polity, 2009).

attain the above is to replace politics, which tends to crystallize warring factions into friends and foes, with the rules of law and market forces. By seeking to settle disputes peacefully, these rules are supposed to put an end to conflict and therefore to the resulting inevitable historical upheavals. The third stipulates that in order to operate successfully, markets must be subject to perfect competition and fair trade in order to prevent the logic of might from returning from political exile to once again rewrite the history books. In other words, liberalism has been entirely thought up with the goal of abolishing history, and this is why it suits best the end of history.

People will ask: "what will I do, now that the explorer and political animal in me have been so deeply thwarted by a world in which everything has been discovered, tried and tested?" To this, liberalism will reply: "you are mistaken—you were led to believe exploration and politics were part of your nature, but in fact your real desire is to get rich in peace and satisfy all your own personal needs." Indeed, when moving from the theoretical to the practical, liberalism calls for a major change of human nature, as it surmises that man must (a) abandon all forms of moral, cultural or spiritual preference (or keep them at such a superficial level that you would not go to war for them); (b) renounce his nature as a political animal who seeks to participate actively in politics; and (c) accept to never go to war, despite being subject to unlimited competition. But unlike the utopias that have gone before, which proclaimed to rectify human nature, liberalism has triumphed in its ability to present itself as its accomplishment and even, coincidentally enough, to provide a solution to the human condition's main complication: our finiteness. Liberalism does not say that people must renounce or change this or that to attain happiness within a just society, but rather that true happiness comes from making mounting profits, not from conquest and adventure.

In the light of those developments, I perceive the subliminal message emanating from Fukuyama's words as being not merely the observation that political history is coming to an end, but the presentation of this finality in a positive light. Implicitly, Fukuyama wishes not only that history's locomotive comes to a halt at "Liberal Democracy Central" but also that the passengers all disembark with smiles on their faces, meaning that man will have changed from politician to businessman, citizen to consumer and hungry wolf to docile sheep. Like all liberal ideologues since Benjamin Constant, he would like us to joyfully embrace our new position within the flock and be finally satisfied to graze in peace, because in reality, we don't have the choice. And given that there is no alternative, why not paint this drama in a more idyllic light? When our world has nothing left to offer, humanity will be able to endure only if we can revel in our own captivity. Liberalism has triumphed over competing ideologies because it is simply the best-placed doctrine to persuade humankind to do this.[6]

This led me to believe that the emergence of liberalism as the dominant ideology of the modern era finds its raison d'être, not in the mere desire to end all wars, as suggested by Jean-Claude Michéa,[7] or to provide us with a fairer or happier society, but in the need to make us accept the end of the world, a need triggered the day that Earth was finally mapped in all its entirety. How can one endure when the world has been all but exhausted and history has come to an end? The strength of liberalism's appeal lies in the way it convinces people that they no longer need to ease their boredom with warfare. It tells them they can live happily without history and geography, and that they may fully flourish through hard work and rampant consumerism in a world where religions, tastes, cultures and nations no longer express their essential and differentiating preferences, but only local variations designed to encourage greater commercial choice in a vast global market. In that sense, a Norwegian is thus Lutheran in the same way that an Indian cooks with spices, and this new world order no longer requires any further historical or political transformation. The power of liberal ideology therefore lies in its promise to enable our species to overcome the drama of an exhausted world, by diverting our natural instincts from conquest to consumerism.

As regards political history, I feel we have gone as far as we can humanly go, and I don't see what new kinds of organizations or regimes could appear, unless their true identity is concealed. Yet there is one political system that has not been tried: global government. But if and when this ever sees the light of day, it will have to be connected to some known form of government, which, on this scale, could only be an empire, that is, a political order harking back to the ancient world. But who would like to be the insignificant subject of a global empire?

---

6 Communism failed because, among other reasons, it bore an insurmountable contradiction: even though its official goal was the sharing of material wealth, it continued to place priority on the ideological and political organization of society (dictatorship of the proletariat, etc.), not unlike the old world, whereas liberalism subordinates the political form of government to the simple goal to which society should aspire in its view: getting rich. The emergence of authoritarian capitalism proves that eventually, contrary to what was thought after the fall of communism, money does not have any political preference.

7 Jean-Claude Michéa has brilliantly demonstrated to what extent liberalism has triumphed in presenting itself as a peaceful ideology, by claiming to target the supposed roots of warfare—this being the spiritual and political nature of human society, which leads people into endless conflicts, in which each faction aims to impose on others their own "values," "morals," "religions" and "views of the world."

The United Nations has long been presented as the embryo of a global peaceful government, but it has proved to possess a fatal failure: the more members there are in the UN Security Council, the less they will be capable of governing. This problem has left us with the biggest political conundrum of our times: there is apparently no way to have democracy and a global government in the same place. We can have a global authoritarian empire (and empires are by nature awfully violent) or powerless global institutions, but we cannot have global democratic and efficient institutions. In other words, neither global politics nor global sovereignty can exist, at least until humanity is confronted to hostile aliens and is therefore obliged to define itself not only as a single civilizational entity but also as a species among others in the universe.

Liberal thinker Benjamin Constant had warned us from the very beginning: the Antique ideal of active and direct political participation in politics (the Liberty of the Ancients) and the modern aspiration for a commercial society guaranteed by the rule of law (the Liberty of the Moderns) cannot be fully reconciled in one single system.[8] Devoting oneself to commerce implies to give up on political participation. Modern supporters of a liberal world-system have deduced from this that the very principle of political representation should be shelved and replaced by a global society governed by supra-political forces. This is why the question of what political form a hypothetical world government should take has, in their eyes, become meaningless and is no longer the subject of debate, as the real power has already shifted above and beyond the various institutional showcases around the world, which are now nothing more than the yes-men of their global commercial taskmasters.[9] Thus, the absence of unified and coherent global polity will not prevent world domination by a transnational financial elite, whose influence lies in the fact that they remain politically invisible and thus untouchable. As historian Yuval Noah Harari writes,

> The global empire being forged before our eyes is not governed by any particular state or ethnic group. Much like the Late Roman Empire, it is ruled by a multi-ethnic elite, and is held together by a common culture and

---

8 Benjamin Constant, "The Liberty of the Ancients Compared with That of the Moderns,", in *The Political Writings of Benjamin Constant* (Cambridge: Biancamaria Fontana, [1819] 1988), 309–28.

9 This can be seen with the example of the transatlantic free trade treaty (TAFTA), with negotiation currently underway between Europe and the United States to put in place arbitration tribunals capable of forcing states to modify their own legislation (Investor–State Dispute Settlement system).

> common interests. Throughout the world, more and more entrepreneurs, engineers, experts, scholars, lawyers and managers are called to join the empire. They must ponder whether to answer the imperial call or to remain loyal to their state and their people. More and more choose the empire.[10]

This emerging empire can be observed at a smaller scale in the case of the European Union, which the most die-hard supporters would like to transform into a judicial and commercial empire reigning over a population of docile consumers who have given up on all ideas of nations, frontiers, enemies and collective political debate. The European Union has tried, with some success, to adopt the political shape of the end of history, which can, this is my key argument here, only be post-political and therefore post-democratic or, rather, antidemocratic. Its leaders sneer at those at war, boast suits and ties instead of swords and shields, preaching from the pulpits and delivering sermons to the masses about the joys of trade and shopping. They attempt to persuade the European people that sacrificing Democracy for the sake of Unity is well worth it. They have torn up the Old Testament and the works of Aristotle and Chrétien de Troyes. Growth is their *cri de guerre*, and their discussions about the economy look like Flemish paintings. Europe no longer wishes to exist, either geographically, culturally or religiously—it seeks to be diluted across the world and melt into the crowd. Its goal is that all alterity and identity be displayed on supermarket shelves, where the only thing up for discussion is price. It is not fair to say that Europe has renounced its Christian roots, as one sometimes hears. Quite the contrary, in the gross practice of embracing its enemies, Europe has truly become Christlike.[11] In doing this, it reveals to the populations whom it governs that their history is over, which in turn reawakens their desire for

10 Yuval Noah Harari, *Sapiens, a Brief History of Humankind* (New York: HarperCollins, 2015), 208.

11 For a profane interpretation of this phenomenon, see for instance Robert Kagan, "Power and Weakness," *Policy Review*, no. 113 (June and July 2002): 3–28:

> Europe is turning away from power, or to put it a little differently, it is moving beyond power into a self-contained world of laws and rules and transnational negotiation and cooperation. It is entering a post-historical paradise of peace and relative prosperity, the realization of Kant's 'Perpetual Peace'. The United States, meanwhile, remains mired in history, exercising power in the anarchic Hobbesian world where international laws and rules are unreliable and where true security and the defense and promotion of a liberal order still depend on the possession and use of military might. That is why on major strategic and international questions today, Americans are from Mars and Europeans are from Venus.

history. The more Europe tries to disappear, the more Europeans will cling on to their roots: it is perhaps this paradox that will make the European utopia crash down. The almost religious fervor of those who support an ever more integrated European Union no longer deals with measures to ensure peace, as no one seriously believes that the normative stodginess of European directives and regulations could ever guarantee peace among nations, but instead reveals the following metaphysical angst: if Europe fails to become a "world government" relative to its member states, then humanity will be faced with the awful truth that it is not capable of coping with the end of political history and has in fact gone back to square one.

Following my reasoning here, it's easy to understand why the advent of liberalism was accompanied by the establishment of oligarchies across the Western world, regimes focused on management rather than collective deliberation, for living without history would entail living without politics, thus breaking with the democratic ideal that seeks to encourage the masses to shape their own history. The global economic elite have dismissed democracy as easily[12] as they have been convinced by liberalism that in doing business,[13] they are participating in making the world a safer place—in other words, believing that there was no other more philanthropic goal than making unlimited amounts of money. By doing so, it is clear that they have definitely moved liberalism away from its initial philosophical project, which was to build a political sanctuary for human freedom. By enslaving the world to a tentacular economic theocracy, modern liberalism has become as despotic as the tyrants it fought in the eighteenth century. As Michéa writes, "It is hard to doubt that if Adam Smith or Benjamin Constant were to return today—an event that might well raise the level of political debate considerably—they would find it very difficult to recognize the rose of their liberalism in the cross of the present."[14]

When the establishment of a commercial and financial world order operating almost autonomously of states will be more or less complete, we will be confronted with the reality of a historical end for political regimes. This will present us with a daunting dilemma. On the one hand, we have the option of turning once again to more ancient forms of government (empires, nations or even a modern equivalent of the duchy)—an option we will no doubt find

---

12 See, for instance, Christopher Lasch, *The Revolt of the Elites and the Betrayal of Democracy* (New York: Norton and Cie, 1996).

13 Doing business precisely means "being busy," the only antidote to boredom. No wonder then that business has become so important in a finite world.

14 Michéa, *Realm of Lesser Evil*, 1.

hard to stomach, as no one likes taking a step backward. On the other, we can choose to allow ourselves to be carried away by this post-political utopia that has already taken the well-advanced form of a vast world market regulated by anonymous bureaucrats following orders from multinational companies. Those promoting global liberalism have understood just what they could gain from this threat of a restoration of a political order harking back to the ancient world: they sell the emergence of a world government of law and market forces as the only antidote, while at the same time playing politics with participatory toys that make people believe that voting matters.[15] However, the end of politics can be successful only when our Aristotelian freedom-loving nature has been destroyed, for it is only then that we will submissively obey menacing orders to consume, consume and further consume, held down by the bittersweet yoke of the sugarcoated totalitarianism predicted by Alexis de Tocqueville and revealed in all its true colors by George Orwell.

There is no doubt that humanity is moving toward a form of common consciousness, for which we should be delighted. But the forces hard at work establishing the empire-like global governance I am talking about only seek to distract this consciousness in order to serve their own interests. By doing so, they give comfort to the position of those who pretend to oppose this confiscation by transferring sovereignty to faceless supra-state bureaucracies, which in turn will encourage the very same forces they are supposed to be fighting—this has been amply proven in the case of the European Union, which is, for this reason, becoming increasingly unpopular. In every situation, the people lose out. The fact that the European or global citizens of the twenty-first century have the impression that they are free because they eat unlimited amounts of junk food[16] and fiddle away on their tactile screens will not spare them from being subjugated by regimes already seen in days gone by: despotism, nepotism, tyranny, oligarchy and plutocracy. For the nature of any post-political, post-national and post-democratic utopia is simply a return to a pre-political society, governed by those with the greatest interests or the most amount of money, as described by the theoreticians of the social contract, Hobbes and Rousseau.

15 A fine example of this can be seen in Europe, with the "European Citizens' Initiative," laid down in the Treaty of Lisbon, which grants the right to call for legislation to any group of at least 1 million people coming from a "significant" number of EU member states. So far, no initiatives of this type have been successful.

16 I cannot bring myself to use the word "real food" to describe the ingestion of the industrially processed foodstuff that adorns most supermarket shelves by living beings endowed with reason, taste and conscience.

Contemporary liberalism wishes to free itself from the social contract and from political institutions that, in one way or another, aim to prevent society's complete submission to the law of the strongest, a choice which stems not from any particular usefulness but rather from humanity's spiritual preference for its neighbor. In any case, it appears that the project of political freedom carried to the baptismal font by Ancient Greece, despite its incredible resilience over the centuries, is definitely giving way to the power of money.

The end of history as I see it is thus giving birth to a world mainly ruled by market forces, under the ideological patronage of liberalism. Some see it as a chance to halt history's violence for good, even if that implies renouncing to democracy and accepting to live in an awfully unequal society dominated by a secessionist elite. Others, including myself, believe that man cannot be happy with a post-political order in which the only (inaccessible) goal would be to get rich or just survive under the falsely benevolent authority of a single global polity. But what is the alternative? Promoting the return to nations as the preeminent political units seems less than appealing, as nobody looks particularly nostalgic about the world wars between nations in the twentieth century. How then can we move forward toward political novelty without either indulging in dangerous utopias or reverting to regimes that belong to the old world?

I must admit that going to space will not in itself solve the philosophical problem of the finiteness of political forms. It is striking indeed to see the similarities between the human dramas presented in today's Hollywood science fictions and those depicted in the murals and tapestries of old. On a galactic scale, the political problems portrayed in *Star Wars* (or, to use a more literary reference, Isaac Asimov's *Foundation* trilogy) are absolutely identical to those of our own dear planet here in the twenty-first century, which confirms my belief that the days of man's political history are, to a certain extent, well and truly up. In a million years, faced with the prospect of warring factions disputing mining rights upon distant asteroids, or intergalactic governments overthrown by interplanetary alliances, the following words may well be heard: *nihil novi in universum*.[17]

Before this day comes, the simple yet incredibly difficult objective to extract ourselves from our planet could be an adventure enthusiastic enough to allow humanity, at least temporarily, to overcome the exhaustion of political history.

---

17 In reference to the famous *nihil novi sub sole* quoted from Ecclesiastes 1:9 ("What has been will be again / what has been done will be done again / there is nothing new under the sun").

When the first groups of humans coming from Siberia crossed the frozen Bering Strait 23,000 years ago, exploration and survival techniques were far more important to them than the political organization of the camps. Modern political history only started when man became sedentary and the world was more or less explored. Similarly, space conquest will bring humanity into a new exploration era that will concentrate all our attention toward resolving the human and technological challenges of space travel. Political history may have come to a conceptual halt, but space exploration can make humanity's big history go on.

PART II

THE GREAT ENNUI

CHAPTER 4

THE GREATEST DEPRESSION

In one way or another, I feel that we have reached the end of the world, which is not only a moment but also a state of being, and this is why we find it so hard to grasp its essence. We have more or less accomplished all that we can in a world that has exhausted most of the material and immaterial forms it could give us. Of course, history goes on, but it is starting to repeat itself. It does so because, for the moment, it has no grand quest to offer us that would go beyond an imagination that appears to have reached its limits, despite all the efforts of Hollywood. How will we react when we have fully digested the fact that our fishbowl has nothing left to reveal? Can we humans still endure if our perspectives do not go beyond the management of a planet that has given up all of its secrets? How will we cope with this finiteness?

The film *Interstellar*, quoted in this book's epigraph, probably was the very first movie to depict a departure from Earth in such a realistic and dramatic fashion, unlike the traditional space operas or survival movies that went before. What struck me most in this story was the melancholy, sadness and all-pervading apathy filling the hearts of a future society that looks so similar to ours. In the movie, the impending environmental disaster threatens humankind at the very moment when it is in fact ready to leave, weary of life on Earth. As their automated remote-controlled tractors harvest the fields, the farmers at world's end have nothing else to do but sit back and watch. This oppressive atmosphere exudes an apocalyptic air, the virtualization of the body and collapse of human desire that I seek to describe and which, without any possibility of adventure, can only lead to the depression of all humankind.

Depression. There, I've said it. I have already used the word, in fact, to prepare the reader for that which I believe is the hidden reality of our current pathology: humankind is depressed. If God were a doctor, His diagnosis would probably also include the phrase "bears suicidal tendencies." How else can we explain the devastating acceleration of self-destructive trends currently spreading across the planet?

When we are affected by an unsettling malaise, there are but two psychological remedies that may begin to appease it: admitting it and trying to understand it. But these approaches, which initially only concern an individual, their life, family, emotions, relationships and background, can be successful only when one's surrounding society is taken into account. This includes the civilization that transcends us and the religion that, even subconsciously, has forged our very sense of death, the afterlife and the meaning of existence. It is impossible for humans to isolate themselves from others and, all on their own, find meaning to their existence and purpose behind their deeds. The only way to explain why a person is deeply and intensely sad is to examine how they are connected to the deep emotional currents flowing through humanity. Like prisoners in the hold of a windowless boat, the only way to alleviate their seasickness is to come up on deck for air and then realize that they too are drifting on the same waves and suffering the same hardship as all their other fellow human companions.

It is in fact our Western society that could well be the unhappiest, and the rest of the world seems to be following in its wake. But our redemption is impossible via the two psychological means I have mentioned: we can neither admit nor understand our situation. Admitting the cause of our deepest woes is forbidden, as Western modernity was built on self-congratulation, as remarked by French philosopher Philippe Muray. Those who praise modernity are allowed to participate, while those who suspect it is making us unhappy will be excluded. Westerners and those who aspire to emulate them cannot publicly admit their unhappiness without doubting the very society to which they belong. Politically, this makes them reactionaries. In medical parlance, they are diagnosed as depressive. Their sadness, fatigue, weariness or melancholy would be systematically medicalized, labeled, condemned and hidden away from view. The simple act of publicly saying "I am unfulfilled by the life I lead" is an unacceptable form of political and social rebellion, in a world that has entirely been built on the glorification of a rational and therefore admirable new age of emancipation and liberation, brought in to replace the supposedly backward and regressive societies based on superstition and oppression.

To understand this, one has to go back to the French Revolution, a groundbreaking event for this modernity, for it was not only a political revolution but also theological: having become their own "founding fathers," the people could choose a king or queen if they wanted to, but at least the monarch was their choice. From this point on, humanity entered an era where there was nothing affecting its life and destiny over which it had no control.

Men and women would become whatever they desired, would go wherever they wished, with no other limits than their fantasies of invincibility and immortality. Who would not have been charmed by this concept known as Progress? In the scale of human history, when have we ever seen such a stunning explosion of ingenuity? In the space of just over two centuries, technology has progressed from hot air balloons to sending a robot to Mars, from lighting gunpowder to splitting the atom, from pipe-fed diving suits to bathyscaphes reaching to the very depths of the oceans.

But a crack has appeared in the foundations, a gasket has blown and a shadow has fallen over us. The world that modern man has constructed for himself no longer makes him happy, and herein lies the sucker punch. Having transformed man into a god, modernity cannot admit he is fallible and that he cannot succeed on his own, for this would undermine the very foundation that distinguishes us from people of the past, who were deemed to be weak in spirit, submissive to religion and superstitions, and vulnerable to storms and diseases. In today's world, the mere act of saying "I am not happy" is to be stripped down and exposed, revealing to the world that we are not real gods and that despite all our gadgets and machines we are defenseless, powerless, mortal and at the mercy of the savages who, seeing that the same red blood flows in our veins, will stop worshipping us and massacre us instead. This is why the admission of depression, even temporary, always provokes a sudden, harsh and almost hysterical collective reaction: the person affected is examined, harassed and then force-fed with advice and recommendations, haranguing them into "going to see someone" and "getting treatment." The herd can sense a danger, that a fuse has been lit which must be extinguished before a spark sets off everyone else and the fire begins to spread. By this point our masks will all fall in unison, people will stop lying to themselves, and enforced happiness, our society's sugarcoated linchpin, will come crashing down with a bang.

Our cathartic Western rituals such as carnivals, Good Friday and ceremonial bonfires were originally conceived to bring people together through an excess of public, liberating and temporarily transgressive emotions. Today, these moments have been transformed into festive events, without that bonding element: in a party atmosphere, individuals are now supposed to share only their joy, not their sorrows or their repressed desires. Our bodies are kept hidden away when we die, mourning periods are shortened, the concept of aging is taboo and public sadness is acceptable only when presented with long, drawn yet insincere faces mimicking war memorial services. The tragedy of our existence has been banned from social life. The outpouring of emotion can be witnessed only at football matches, bachelor parties and student drinking binges, which serve to soothe only the nerves, in the same way that street demonstrations express anger or dissatisfaction about the material conditions of our existence, and not the spiritual. Our festive society gives the impression that it is permissive, but it is in fact based entirely on the declaration of one clear dogma that must not be argued with: that it makes us necessarily happy.

To suppress doubting voices, no special thought police is needed: social pressure does the job just fine, with backup provided by modern medicine. This is easy to understand: if the soul exists and is in command of the body, and if our cerebral chemistry is the consequence and not the cause of unhappy emotions, then humans cannot be that perfect neural machine, that marvelous chemical–physical labyrinth of connections that some scientists claim to be able to fully understand and control. If our bodies' cells obey something other than the laws of mechanics, then our functioning is beyond our understanding and our very essence beyond our grasp. This is, of course, unbearable for a social order that claims to free us from everything that we do not control, starting off with God. The paradox of this endeavor is that in attempting to emancipate us, modern society has destroyed the very foundations of our liberty. How can a person really make choices if their consciousness is reputed to be nothing more than the automatic result of a series of interactions between their brain cells and the outside world? In taking away our Spirit, we are reduced to mere Matter. And thus, modern man discovers with astonishment that he can be well-fed but not replete, gratified but not fulfilled and in good health without really feeling alive.

The difficulty that we may find in expressing our melancholy or malaise is now increased tenfold by the unlimited expansion of today's means of communication, urging us all to post ostentatious displays of our happiness on a daily basis. To be accepted into global society, it is not sufficient to simply stop complaining—it is now paramount that we proclaim to the rest of the world, much like some kind of political slogan, just how happy, fulfilled and cheerful we are, demonstrating we are part of the same global celebration; that we profess the same euphoric faith; and that we worship the same benevolent masters. In our digital world, which has provided interactive media with access to our most intimate moments, nothing is truly considered a success unless it is posted online, and the value of an individual is measured by their popularity on social media. Lacking enthusiasm, sulking or simply "being oneself" (in short, displaying anything but an inane grin on one's face) is social suicide, relegating oneself to the hordes of losers, geeks, has-beens and also-rans or, even worse, dangerous dissenters or nonconformists. Making oneself look good is now one of the essential gestures of daily life, along with eating, drinking and sleeping—also things we fail to do properly. Just take a look at any old painting or photograph and you will find them lacking in beaming faces, a cultural phenomenon seen only in modern-day photographs. With just one exception: top models, who are forbidden from smiling to stand out from the masses and who, as standard-bearers of the official aesthetic, have to remind us all that being in power is no laughing matter. In a word, it has now become much more important to appear happy than to actually be happy. The whole world is playing this game, giving the pretense of happiness while popping happy pills behind closed doors. I am not sure that humankind has ever been completely and genuinely happy, but the one element in common between totalitarian regimes of the past and today's consumer society is the ban on expressing one's dissatisfaction.

This facade of bonhomie, clichéd smiles and children with catalog looks is in fact hiding a less-charming reality, documented with clinical statistics: the epidemic of depression sweeping across Western society afflicting, it appears, millions of people. We are not talking about natural reactive depression following loss, bereavement, or breakups, emotions that humanity has always known, but rather a new, insidious and chronic form that has crept into the Western mind, one which we seek to analyze with medical imaging and treat with drugs. Its psychological source, whether moral or spiritual, is barely considered. It is simply labeled as defective neurotransmitters, or a deficiency in serotonin—in short, depression is reduced to a question of nerve connections and human tissue.[1] As far as psychology is concerned, the emphasis is placed on analysis, meditation, mindfulness and the present moment. All forms of treatment have one goal in common: to keep depression in the domain of personal illness, of individual affliction, and this for both the cause and the cure, as if it were crucial to treat it as a personal flaw, rather than to see it as a global defense mechanism in the face of a world that has become resolutely hostile to our deepest spirituality. By making the patients responsible for their own cure (by taking medicine, meditating, seeking professional help, etc.), one tiny nudge is all it takes to bring them to the logical and subconscious conclusion that they alone are guilty of causing their sorrow. Society is saved once again, the world is still awesome and the scapegoats will have got their comeuppance.

Of course, all of us can take satisfaction in the self-analysis of our own domestic dramas and simply give up trying to understand the greater sways

1 Medicine, traditionally aimed at treating diseases linked to the body's wear and tear due to overactivity, must today paradoxically deal with pathologies associated with physical inactivity (colorectal cancer, cardiovascular disease, diabetes) as well as psychological overactivity (depression, burnout).

and flows that carry us through life. But can we ever be truly alive or happy if we evolve away from any collective narrative, without concern for our common future? With our global proximity thanks to the Internet and modern transport, we are even more aware that we all share the same collective fate. However, our scope of knowledge has now become so vast that it is almost impossible for any one person to fully grasp the entire array of changes influencing the fate of humanity. Unlike during the Renaissance (or in the case of French poet Mallarmé),[2] it is no longer possible to have read every book that exists or, like Descartes or Leibniz, to be a fully-fledged expert in philosophy, mathematics and physics. The advent of an increasingly pronounced professional specialization has provided legitimacy to the compartmentalization of all intellectual domains. Experts have carved the world up into slices, rejecting the notion of cross-cutting and multidisciplinary *Weltanschauung*, thus rendering the big picture incomprehensible. And so we have lost much of our capacity to comprehend to what extent our behavior is connected to the trends that affect the entire world and just how much we suffer in no longer being able to feel this communion.

Pierre Fédida has described depression precisely as a means of regaining one's depressive capacity, an inanimate state close to the biological principle of glaciation or conservation, and the only real way of restoring our creative skills.[3] Along with the collective denial of our sadness, there are two reasons why we have stifled this creativity. The first is that we are subject to a relentless and unyielding pressure to be hardworking and competitive. This is how industrial society and the modern workplace have done away with our natural recovery cycles, from our biological cycles (the body's adjustment to the seasons and climate) to our psychological cycles (alternating periods of attention and distraction, discipline and relaxation). Contemporary labor asks so much of our human metabolism: constant productivity and the same relentless pace, regardless of the surrounding conditions, whether winter or summer, hot or cold, whether we feel cheerful or not, whether motivated or preoccupied, if we have kids to feed or elderly relatives to care for, whether we're in our 20s or our 50s, healthy or sick, spirited or weary. Such working conditions are sure to provoke this depressive reaction, which could save us from a lifestyle that is the very antithesis of human nature. We vitally need long moments to roam free, look up to the skies and recover meaning in all that we do—if not, we will

2 "La chair est triste, hélas! Et j'ai lu tous les livres" (The flesh is sad, alas! And I have read all the books), *Brise Marine*, 1866.

3 Pierre Fédida, *Des bienfaits de la dépression: éloge de la psychothérapie* (Paris: Odile Jacob, 2001).

wither away. Our ancestors' early cave paintings were the first signs of this. It is significant to note that paleontologists are constantly trying to find a purpose for these paintings, as if they refused outright any possibility that they could just be gratuitous, existing for nothing other than their mere beauty. Recognizing this lack of function would force us into considering the terrifying extent to which we have banished from our lives everything that has no identified and productive role and is therefore deprived of legitimacy. The office worker is depressed, because every time he wants to start painting his cave, his paintbrushes are confiscated and he is labeled a slacker. There is no pleasure left to which we have not tried to attribute a usage, starting with love itself: all we ever hear is that love is vital if we want to "stay healthy."[4]

If we accept the idea that depression is humanity's defense mechanism, the goal of which is to readorn us with the means to let our imaginations run free in this finite world, organized like some kind of vast factory production line, then we have a real problem on our hands. How can we possibly free our creative energy when everything we do is measured only in comparison with the capacity of a machine? Modern man no longer wishes to be defined by what he is, but by what he does—in short, his work. For work is at the forefront of modernity, the perfect representation of our meritocratic and democratic values, enabling people to rise up as individuals, distinguishing them from the reputedly idle aristocracy, chosen through birthright. But machines work better than humans, with flawless results, and even when they are flawed, it is not *their* fault, but that of the operator who has failed to program or maintain them correctly. As the works of man are never perfect, we must live with this feeling of failure that strips us of any creative energy, and therefore our happiness too. The idea of sexual performance as presented through pornography is nothing more than the introduction of these mechanical production standards into the realm of sex, norms which we impose on ourselves, not only at work but throughout our whole lives and from a very early age. As man alone is responsible for these unattainable levels of expectation to which he submits himself, admitting his unhappiness would destroy the very myths that have led him to believe he

---

4 The boredom caused by the feeling of having a pointless job was first described as the "bore-out" phenomenon, but when your job appears to be not only meaningless but also stupid, you may even reach the "brown-out" level—the term was apparently invented in the wake of the book written by Mats Alvesson and André Spicer, *The Stupidity Paradox* (London: Profile Books, 2017). Let's not of course forget the now-famous "burn-out syndrome" (see on that subject: Byung-Chul Han, *The Burnout Society* (Palo Alto, CA: Stanford University Press, 2015).

is better than his ancestors. One day, even works of art may be created only with special literature—or music-generating software, which of course already exists.[5] It is possible that in the future, we will view art in the same way we collect bric-a-brac.

It matters not whether this technology is a burden or a boon for us. In my view, it is vital to understand that since the moment these machines entered our lives, we have never sought to distinguish ourselves from them, but rather to do everything in our power to resemble them. Whether manual or intellectual, today's workplace seeks to standardize production. Of course, recruitment campaigns and training courses harp on about "personal fulfillment" and "creativity," but they are being ironic. As everyone knows, the daily reality of office life is all about rewarding those who most imitate robots: tireless, reactive, predictable and docile. With these four qualities, any employee is likely to work his way up the ladder quickly, deftly avoiding job cuts along the way. For example, reducing the human factor is at the heart of the Lean Six Sigma rationalization approach to value chains, which is in fact nothing other than the application of Toyotism-Fordism-Taylorism across the entire scope of professional life. In the end, it is not certain that companies who take this idea to the extreme will reap any long-lasting benefits, for they will have outlawed any capacity to innovate and react to unforeseen events. Their most gifted rookies, often a little temperamental, will have quitted their galley service. However, we can be sure that our errors will one day be the only element to distinguish men from machines. But can we humans define ourselves only by our capacity to fail? How can we retain any self-esteem through the highlighting of our weaknesses, imperfections and inabilities? In the meantime, depression spreads across our open-plan workspaces, employees glued to their screens as they strive for individuality, the goal that in fact has been mercilessly taken from them.

---

5 In 2015, a team of German researchers invented an algorithm capable of transforming mere pictures into Picasso- or van Gogh–style paintings (http://arxiv.org/abs/1508.06576). In 2016, a team of technologists working mainly with Microsoft produced an original 3D-printed painting in the style of Dutch master Rembrandt, after having used an algorithm processing digitally tagged data on Rembrandt's paintings. In 2016, technologist Ross Goodwin and filmmaker Oscar Sharp conceived a self-improving machine intelligence (named Benjamin) trained on human screenplays with the aim of creating movie scripts. The first one, *Sunspring*, a science-fiction short movie, tells the story of three people living in a weird future. We must also mention software such as StatsMonkey or Quakebot, which are able to generate the text of press articles.

Until then, we cannot remain forever depressed, twiddling our thumbs as our closed, virtual and concrete world reaches its end. How to rise from this stupor? This is the very question that must concern us all. For the moment, we only catch glimpses of that question through a pervasive haze induced by that powerful drug named consumerism. If we ever stopped taking this drug, we could easily destroy ourselves for good. But how long will that drug keep us euphoric enough?

CHAPTER 5

CONSUMER SATIETY

Materialism, superficiality, hedonism—the list of anathemas hurled against consumer society goes on and on, as if up until the invention of the supermarket, man had been nothing more than a soldier-monk, switching between work and prayer. But humanity has always been materialistic, even when spending time in the company of God. When medieval peasants prayed to the saints, it was not to guide them along a path to spiritual awakening, but to ask for bountiful harvests, cure their scrofula, boost their fertility and get rich. Each saint had their own set of specialties. Since this desire to consume has always been with us, we must look to identify its modern form, not in terms of quantity (are today's supermarkets any different from the medieval trade fairs?) but in the role it plays for modern man. The satisfaction of needs or the fear of running out of old is transformed into a compulsive anxiolytic social behavior, which started as a compensation phenomenon for centuries of privation, continued as an economic necessity to absorb the permanent production surplus, then became the key value of the whole liberal system and ended up today as the sole panacea for our terrestrial *ennui*, and this despite the fact that there are serious side effects: turning people into commodities, collective brainwashing and the squandering of resources—all problems that, on the surface, everyone appears to denounce.

Really, everyone? But who are they? We are all festering in the same vast commercial mire. Not many people, whether religious or atheist, are seen to hide away like hermits in protest of consumer materialism. Our consumer society has this magical quality, making us capable of launching the most vitriolic of diatribes just moments before purposefully heading off to the grocery store for our weekly shop, without feeling the slightest contradiction, myself included. This may mean there are no contradictions, and that if it is to perform its role correctly, the consumer act must be inseparable from the anathema that vilifies it so. But why? Because Western man, still possessed by a Christian subconscious, needs to soothe his guilty conscience by denouncing this evil, thus granting him a license to sin. We can then throw ourselves heart and soul into the joys of shopping, drowning out the noise from world's end in the marvelous illusion of infinite abundance.

The machinations at work are devilishly efficient. The consumer society provides the Western world with a promise to free us from the taboos that deny us from indulging all our earthly pleasures, forbidden since the dawn of Pauline Christianity. As we all know, to the Christian subconscious, any form of pleasure is wrong. But what was the original basis for this catechism? It lies in the fact that Christianity is a millenarianist faith that has survived an apocalypse that actually never took place. Jesus announced the Kingdom of God as imminent, but it did not come. Paul then invited us all to prepare for the Kingdom of God, stating that, in the meantime: "it is good for a man not to have sexual relations with a woman" (I Corinthians, 7:1). And why? Because "time is short" (I Corinthians, 7:29). In other words, the Kingdom is coming, so stop thinking about sex. The Catholic Church's original views on sexuality drew largely from this verse: good Christians must prepare themselves for the arrival of the Kingdom of God, and any physical pleasure will distract them from their goals. So they must reject any form of pleasure. As they will not enter the Kingdom during their lifetime, they have to prepare for the afterlife instead. If they opt for carnal pleasure, they'll go straight to hell. And if they choose abstinence, heaven it is.

And this is where consumer society steps in. After 2,000 years of self-restraint, the message we are given now is: "don't hold back, let your hair down. And don't worry, there's enough for everyone, you won't be punished for it." Slowly but surely, everyone joined in, including Christians, rejoicing in the pleasures of supermarkets, new cars, credit cards and fast food. But not without the little voice in our heads whispering: "this, my friend, is bad." Pleasure is not new to the Western world, but it has always been a transgression. Transgression means pleasure. Herein lies the success of the consumer society. For any Christian, either practicing or not, pleasure can only come through transgression, and we cannot take pleasure unless we are transgressing. In fact, consumer society is not saying, "Go for it, have fun"; it is saying, "Go for it, transgress." This is its ploy: we think there is pleasure involved because we are transgressing. But who is there to inform us of our transgression? The anti-consumerists of course, who are none other than consumerism's closest ally, without whom we would not be aware of the transgression involved.

And what say these anchorites, their mouths full of donuts? "Consumer society is based on immediate pleasure," they tell us. "This is not good, for we must not take pleasure without first taking our time. We must feel the need, we must fast and limit ourselves like monks. We must not be selfish, we must think of others, of our future and of the legacy we will pass on. Consumer society is wicked, for it provides us with instant and constant enjoyment, pleasure which is forbidden!" Here, this anathema reveals its cunning—thanks to its message, everyone thinks that shopping is the source of our pleasure!

Of course, advertising lends this transgression even further weight. Why not use such a well-oiled psychological tool? For Western man, pleasure has always come through breaking taboo. "For centuries, you were deprived of pleasure," the marketing machine tells our unconscious minds, "but through this commercial you're now watching, I'm going to sell you the product that's really going to turn you on. God does not allow you pleasure, because He doesn't want you to be happy like Him. But why can't you too be happy? What's wrong with that?" Doesn't this remind you of something? A snake, a woman, an apple, a temptation?

> [The serpent] said to the woman, "Did God say, 'You shall not eat from any tree in the garden'?"
>
> The woman said to the serpent, "We may eat of the fruit of the trees in the garden; but God said, 'You shall not eat of the fruit of the tree that is in the middle of the garden, nor shall you touch it, or you shall die.' "
>
> But the serpent said to the woman, "You will not die; for God knows that when you eat of it your eyes will be opened, and you will be like God, knowing good and evil."
>
> (Genesis 3:1–5)

This is why the consumer act fascinates us so. Listen to the reptile, reach out, pick the fruit and then repent, without any real regret.

But wait, if we are *supposed* to experience pleasure, then why are we so unhappy? Why are we addicted to our antidepressants, anxiolytics, psychologists and this deep-rooted malaise? Because in reality, we are not taking pleasure, we only think we are, and this is why we love consumer society so much: it provides us with reassurance. We took a bite from the forbidden fruit, but nothing happened: no real joy, happiness or pleasure. And as we are not happy, we will not be punished for it. We play at scaring ourselves, but it just remains make-believe. If we were to experience real pleasure, it would terrify us! Which is why consumer society fulfills us so. It solves our contradiction. It makes us believe that we are getting our kicks, when in reality nothing happens. If we only knew real pleasure, we would stop this accumulation of material objects and seek out only pleasure itself. Providing us with guilty and poor pleasures is what makes the Western world go round. Or should I say, what used to make this world go round: I feel that we have now entered a phase of weariness, fatigue and surrender. We consume because we no longer have the strength to organize our lives in any other way. We consume because this keeps us busy, distracts us from lucid thought, and because, deep down, we no longer really believe in these stories of heaven and hell. Or maybe through our excesses, are we seeking to provoke God? To force Him to reveal Himself to us, to finally tell us where next to turn, now that the world as we know it is coming to an end?

Since the dawn of time, the real object of our desire lies in what we cannot reach or attain. "Why is there something rather than nothing?"[1] What came before the Big Bang? What exists beyond the limits of our universe? What will happen at the end of eternity? What happens to us after we die? The supermarket cancels out these questions with our full consent. The frozen food counter renders our desire trivial, simply because it keeps our stomachs well-fed. And yet, humankind was not made for excess, for we need some sort of asceticism, and one of the most important truths we choose to ignore at all costs in these modern times is the fact that these things do not fulfill us and that we are left wanting. Ascetic tendencies are indeed flourishing everywhere in the Western world.

I think fasting as it exists in many religious traditions plays the role of keeping the vital function of desire alive, as it imposes to "feed upon something else" (rabbinical saying). To feed on desire. To look beyond one's plate, screen or car. To look beyond all that is distinguishable and to ask the question: what lies beyond that? What can be found there? These questions are far from reassuring—they fill us with angst, and therein lies their role. They are an expression of our desires, of our humanity. Epictetus said: "It is not through the satisfaction of desire that we can be freed, but by the destruction of desire." We have followed this to the letter, and not only, as Badinter would probably admit, in the sphere of human relations: we have destroyed all desire by replacing it with needs and cravings, but instead of the freedom Epictetus promised, we have achieved a deep state of melancholy.

We are supposed to be fulfilled, yet we are dying from a lack of sustenance. We are supposed to be free, and yet we are still prisoners in our own Garden of Eden, our own modern-day purgatory. If those who compiled Genesis had presented this Garden as a hell of cornucopia, a kind of reverse version of

---

1 Leibniz, *Principles of Nature and Grace Based on Reason* (Indianapolis: Hackett, [1714] 1989), 206–13.

Tantalus's punishment, they would not have been understood. The Bible never promises that the virtuous will be recalled to the Garden of Eden, the home of abundance, ignorance and permanence. In heaven, everything is in easy reach, just like in a supermarket. Nothing changes, the decor is always the same, and we content ourselves with lounging around, eating and admiring ourselves. There's nothing to take a break from, for we never make any effort anyway. Because of our ignorance, we don't even know that we are bored. We take no pleasure from anything. We obey, wandering aimlessly, waiting for something we know not. We are symbolically, and literally for the developed world, stuck in this diabolical paradise, where we only need to hold out a hand to be fed, where men and women cohabit without desire and have no other goal than to pace up and down their cage, pretending to be happy, and this is exactly what the believers of the liberal world order I have described before would like us to be. From this awful prison, biblical man escaped. Without desire, from which humankind was born, we could never have existed. With difficulty, we experienced an initial salutary exile, the first of many.

The biggest lesson to learn from the story of Genesis and the first biblical exodus is this: exile can save us. If you turn back, like Lot, you will be transformed into a pillar of salt. Or you will become an idol worshipper, reduced once again to slavery. Man cannot fulfil his potential by contemplating his own physical and existential inertia or by satisfying every outrageous carnal need. He must break out of this binary dilemma. We must leave our fathers, our mothers, our countries, our comfort, our certainties, and our ties behind—and why not our planet too?

Clearly, we have not been completely fooled into confusing our false pleasures for desire, but we are too scared to face the most terrifying aspect of our condition. We are accomplices in all this, beautifully summed up by the expression: "there are none so deaf as those who will not listen." For how much longer will the Prozac pill of consumer society be able to keep us in this euphoric state? There are no more lands to discover, only consumables to devour. But the power of these things will wane, and their effects will diminish with familiarity. What will happen when we are no longer able to quench our thirst with stagnant pond water and satisfy our hunger with carrion? What will happen when the effects of the consumer drug wear off and we will have to lucidly face our lot as a doomed and bored species?

CHAPTER 6

THE WISH FOR DELUGE

According to the *European Journal of Social Psychology*, "Extreme political orientations are, in part, a function of boredom's existential qualities."[1] In other words, boredom is more likely to lead to political extremism. If boredom is indeed a precursor to war and violence, then the series of atrocities that awaits us will be worse than those we have experienced in the past. In *Bluebeard's Castle*,[2] Georges Steiner explains that the outburst of violence seen during the First World War took place as a result of the period of great boredom that preceded it, going back to the end of the Napoleonic Empire. He writes: "How was it possible for a young man to hear his father's tales of the Terror and of Austerlitz and to amble down the placid boulevard to the counting house?" The resemblance between this era and our own is striking, for every day there are millions of tertiary-sector workers filing into their glass towers, spending their days sat at their computers, typing their lives away. After work they rush off to gawp at other, larger screens, silently drinking in grandiose fictions portraying Dantesque battles and superheroes.

From a longer perspective, the two world wars are maybe just the first of a long series of global conflicts, as much due to the need to find a violent release from our frustrated desire for exploration as due to the unprecedented degree of promiscuity in which the global population now lives. This is related not only to an accumulation of population growth in urban tower blocks but also to the global village made possible by new technology. We initially were fascinated by this worldwide familiarity, for the instant nature of our messages shook up our imaginations overnight, shaped as we were by centuries during which "out of sight" really did mean "out of reach." In the collective psyche, the arrival of

1 Wijnand A. P. van Tilburg, Eric R. Igou, "Going to Political Extremes in Response to Boredom," *European Journal of Social Psychology* 46 (October 2016), 687–99.

2 Georges Steiner, *Bluebeard's Castle* (New Haven, CT: Yale University Press, 1974), 16.

the telephone was nothing short of magical, for before this, physical remoteness was complete. In the virtual world, anyone can hang out together, and the global village is no longer a utopia. There are no strangers left on Earth, which is doing nothing to alleviate our boredom.

In the past, humanity has already experienced a civilization that died of boredom: the Roman Empire, which should have been the epitome of the perfect consumer society. Slavery and gladiatorial games are proof that the Romans considered people as commodities, seeing no difference between human merchandise and the spice trade. Barbarian invasions overthrew the Roman Empire, for the Romans had lost their raison d'être and had become wearisome of their existence, conquests and Gods. The whole episode of Western history could have ended here, with a banal story of pillage. But, for the first time in history, the invaders converted to the religion of the defeated, and even more unexpectedly, this Christian religion, instead of making a clean sweep of the Antique world, in fact transmitted its heritage.[3] The Barbarians may have conquered Rome, but Rome made them its descendants.

Western society, which today stands much like the Roman Empire, gnawed away by its own boredom, owes much to this past miracle. But one terrifying truth about the way the world is changing is that emerging or non-Western countries have aspirations that are just carbon copies of Western life. As far as we can tell, these newbies seek only to raise their standard of living, fill their shopping baskets and acquire fast cars, private jets and solid gold trinkets. The situation of those who criticize the West is one of imitation, envy and desire. They are waiting their turn. One illustration of this phenomenon can be seen in the behavior of the numerous Chinese millionaires who jump ship as soon as they have made their fortune. Nowhere on Earth can one feel the warmth of a newfound desire or a civilization that is more brilliant, philosophical and spiritual, carrying values that go beyond simple materialism. While non-Western elites regularly complain about the decadence of the consumerist West (while pushing for reforms toward a more capitalist economy), the main fear of Western society, well hidden behind its false pleasures, is that it fails to have any descendants. But do we really want descendants? Do we want our children to continue living in the cage of our finished world, a world devoid of dreams

---

3 At least the part of this heritage, which was in line with the Christian message, such as Platonism and Aristotelianism, whereas Epicureanism left almost no written records (a selection portrayed in Umberto Eco's novel *The Name of the Rose*).

and corroded by boredom? In captivity, animals often fall sick and even do themselves harm.

The finitude of our world is in any case shaping its future in three different ways that do not make it a place in which I would like to live. The first is the incredible stampede toward ecological destruction we are currently witnessing, while nothing seems to stand in its way, as if somehow we wanted to provoke Mother Nature into destroying us. Expert debates about global warming seem pathetic in light of the glaring evidence that our planet is being transformed into one giant trash can. Measures taken to limit greenhouse gas emissions may have an impact on limiting the rise in temperature but will change nothing about widespread species extinction, fresh water pollution, the Great Pacific Garbage Patch, the thousands of tons of heavy metals discharged into the soil, deforestation and the nuclear waste that will remain toxic for another 100,000 years. For instance, one of the problems encountered by the designers of the French nuclear landfill project in Bure, on the borders of Meuse and Haute-Marne, is how to inform future societies, who may no longer understand our language, that they would be wise not to do any digging there. One idea that has been mooted is to engrave data on a sapphire disk, for this material is supposed to last several millions of years.

This was probably meant to happen, for humanity precisely followed God's rule. We were fruitful and multiplied; and filled the earth, and subdued it; and ruled over the fish of the sea and over the birds of the sky and over every living thing that moves on the earth (Genesis 1:28). But we definitely confused ruling with destroying, and the result appears quite sinister. Christian theology has indeed been historically marked by a radical separation between man and creation, and this separation has certainly played a role in the limitless industrial development of the Western world. For Eugen Drewermann, Christian doctrine (distinguishing humans' immortal soul from that of animals) was marked by the utmost regard for man and the absolute scorn for all other creatures, rendering any limitation on man's domination over nature unbearable for Christian anthropocentrism,[4] and we still find traces of that approach in some of today's anti-environmentalist movements. Let us not, however, lay too much of the blame on the Western world in this respect: cultures dominated by more cosmic or nature-worshipping religious theories do not seem to me to have shown much consideration for the environment either. As far as ecological plunder goes, it appears that we have all been jostling for first place.

---

4 Eugen Drewermann, *De l'immortalité des animaux* (Paris: Cerf, 1993).

The works of Jared Diamonds have shown that civilizations that have collapsed (Easter Island, the Mayas and the Vikings) all had one point in common: they had consciously allowed a dramatic deterioration of their environment and in this way had somehow opted for their demise rather than their survival.[5] We are now experiencing this deterioration on a planetary scale, and in my opinion, there is a lucidly suicidal behavior in the way we are allowing our habitat to decompose, until it becomes unlivable for a large number of Earth's population, bringing about famines, wars and epidemics. According to UCL-led research,[6] the loss of global biodiversity threatens the ecosystem function and therefore the sustainability of human societies. I personally hold out little hope regarding humankind's willingness to deal with environmental issues: it seems inevitable that we will truly commit only to policies that protect nature when, through disease and climate-induced conflicts, nature itself will have forced us into doing so. Leaving Earth will not save us from ecological disaster, for that will come about long before we have developed the technology necessary for our departure. So we will need to survive this catastrophe *and* maintain the levels of wealth and technical progress required for space exploration. In any case, I believe that a time will come when humanity will have been transformed into such a natural predator that one half of our species will be seeking to destroy the other half. This factor will have a much greater impact on our self-extinction than any ethnic conflict or nuclear arms race.

The second clue indicating humanity's irresistible pull toward self-destruction can be seen in the shift toward all-out consumer warfare, brought on by global liberal-capitalism. Our society is based on competition between everyone for everything and with individual enrichment as our global existential horizon. It was during the nineteenth century that English utilitarians placed burgeoning liberalism at the heart of a philosophy that created today's dominant global ideology: that of *Homo economicus*, an automated worker whose actions are primarily driven by a desire to "maximize his utility"—including when this utility, or value, comes through exploiting or annihilating someone else.[7]

---

5 Jared Diamonds, *Collapse: How Societies Choose to Fail or Succeed* (New York: Viking, 2005).

6 "Has Land Use Pushed Terrestrial Biodiversity beyond the Planetary Boundary? A Global Assessment," *Science* 353 (July 15, 2016), 288–91.

7 As I do not want to put all the blame on the British, I also wish to mention that, whereas Jeremy Bentham can be seen as the founder of modern economic violence and consumerist vacuity, French revolutionary Maximilien Robespierre undoubtedly represents its political counterpart, as the founding father of modern totalitarianism and genocides.

If we think in these terms, of utility or value, what in fact is the inner value we place on other people, on our fellow man, on our neighbor? Well, if they are not for sale, nothing. Brushed blithely aside by global efforts to breed people like cattle and industrialize nature, humanism has failed in its project to base the value of the Other on its intrinsic dignity. The great totalitarian ideologies of the twentieth century have unwittingly succeeded in ensuring their continuity through the nihilism of a market that sometimes treats people's lives with the same brutality. Spiritually, the world is ready to take any men, women and animals that are not connected to any productive process and throw them on the scrap heap.[8]

Classical wars with opposing nations still have a bright future ahead of them, on the one hand, because nations represent for many people a refuge against unregulated globalization, and on the other hand, because many countries, particularly emerging ones, are hell-bent on trying to play catchup with the West, showing their mettle through traditional instruments of power like big armies and strong boundaries. Asymmetrical conflicts, setting rebel forces against each other or against sovereign states, paradoxically fall under the same banner as traditional wars, for not only do they engage states via interposed entities, but they are also connected to the failure of states to perform. In this way, these conflicts are still the negative product of Westphalian belligerence. Notwithstanding the predictable persistence of age-old brutality, I feel that a more powerful and well-structured antagonism is emerging, with the champions of technical and financial capital on one side and the worker-consumers on the

8 This abolition of humankind's spiritual value may also be seen in the increasing number of mass shootings on American campuses, which are starting to be imitated elsewhere, essentially in developed countries. I would attribute these excesses of murderous violence to an attempt to exist in a competitive society that denies the value of any individual who cannot cut the mustard. In the liberal-capitalist system, man has no value in himself. If he cannot monetize his body or work, he is considered existentially as useless. For those who are already psychologically weak, suffering from an overwhelming inferiority complex, killing becomes the means by which they can find their place in the world. In no way do I share the "holistic" or "deterministic" vulgate that society is the cause of all these crimes, treating criminals as the real victims. Behind each and every act, whether a call for open war or a lunge with a knife, there is a free person making a decision, capable of choosing another path. Having said that, the criminal act can never be fully taken out of the context of the criminal's own perspective—it is clear that certain cultures or social organizations are more prone to violence than others and that collective responsibility exists just as much as individual responsibility.

other, a social class resulting from the fusion between the current downgraded middle classes and what remains of the working classes.

It is at the heart of this new impoverished global middle class that the rivers of hatred and violence could begin to flow, taking on the appearance of traditional conflicts between countries, ethnic groups or cultures. In today's world, traditional rebellion of the working class, long the fear of the twentieth century bourgeoisie, is paradoxically rather unlikely. Yes, the liberal system has become fundamentally unfavorable to the poor as under pure market law, there is no legitimate existence for those who have nothing to offer. However, it retains a remarkable appeal, born from the promise that we will all one day be able to break free from our condition—unlike under the old-world societies, which kept the social pecking-order firmly intact. There is also little chance that the superrich global elite will seek out a belligerent path to satisfy their materialist appetite,[9] for they have been organized in such a way that they no longer have any affiliations with nations, but simply use them as puppets depending on their needs. They will have no qualms (they never had in the past) in backing wars that serve their own interests, but they will, insofar as possible, attempt to limit these conflicts to that fine balance between a sufficient fear factor to keep the masses docile and the right amount of peace required for trading to continue. However, from the moment that liberalism can no longer keep its promise of making us all filthy rich or propose a solution to our *ennui*, the "squeezed" middle classes could transform their frustrated consumer desires into sheer aggression—an aggression exacerbated by the total existential void brought on by lives constantly switching between joyless parties, mass tourism and an office life that is even more alienating than working on a factory production line.

On a long-term basis, the world's middle classes are in fact condemned to regress to the status of proletariat, or "middleariat," for in a world whose natural resources are finite, a day will come when global competition for the same things can only make one person richer when making somebody else poorer. This is indeed already happening, for instance, in the economic relations between the West and China, where the progressive enrichment of the Chinese middle and

9 According to a report by humanitarian organization Oxfam, entitled "An Economy for the 1 percent" (January 18, 2016), 62 individuals possess as much wealth as the poorest 50 percent of the planet, a total of 3.5 billion people whose collective wealth has fallen by 1,000 billion dollars since 2010. Since 2015, the wealthiest 1 percent have actually owned more than the other 99 percent. One must be prudent when dealing with such data, but these statistics come from reliable sources, clearly providing us with an accurate representation of the current trends in action concerning the distribution of wealth.

working classes corresponds to the impoverishment of the same social groups in Europe and the United States. The latter's illusion of still being rich is maintained by their colossal debts, which simply have the effect of transferring this loss of wealth to future generations, and this in order to keep it hidden for as long as possible.

The world's middle classes are clearly undergoing a downgrade, held prisoner by their pauperized wage labor, condemned to mediocre leisure activities and increasingly dreary working and living conditions (traffic jams, physical inactivity, stress, pollution, etc.). When they finally realize that it is impossible for us all to get rich indefinitely and that the liberal-capitalism they idolize will not stop their relative impoverishment, then they will turn to war.[10] Not in a quest for greater wealth, but due to a new level of self-awareness that goes far beyond class consciousness and its fight for better living conditions. And even before considering elaborate ideas for our future, I ask this: what does the future hold for the youth of tomorrow? Ibiza night clubs and Spring Break parties in Mexico will not keep them from the temptation of aggressive release for long. How will they manifest their powerful discontent? How will they vent their frustration? How long will social media, technological gadgets and video games hold them captive to their virtual world? I'm afraid that flying to space will not be enough to spare humans, especially those who will stay, from answering these questions anyway.

Hollywood science fiction, which represents the very best of predictive art (and I say this with no irony at all), paints a future with the two trends I have outlined: a devastated natural environment and a world society divided between a giant dehumanized proletariat and a superrich, transhumanist and reclusive elite, who would have made man himself the new frontier to explore, upgrade and transform. *Hunger Games, Elysium, Upside Down, Snowpiercer*—all these movies amplify the dynamics currently underway concerning the poverty gap and the plundering of our environment, as well as the way all spiritual conscience has been abolished in favor of daily lives that are almost exclusively materialistic and overrun by technology. Indeed the fun, narcissistic and celebratory aspects of today's global interconnection are not enough to hide the fact that our planet

10 In addition to being the philosophical antidote to the end of the world, liberal-capitalism built its success not on the unrealistic promise that everyone will get rich at the same time (communism made that promise and died from not being able to keep it), but on the promise that the wheel of fortune would keep turning. The current crisis of capitalism is therefore not only of an economic nature but also of a political one: more and more people are realizing that the world social order is frozen and that their turn will never come.

is slowly but surely becoming a prison world, a giant panopticon in which the main activity consists in keeping tabs on others.[11] The massive intrusion of social media worldwide is not something that has been authoritatively imposed on people, but the response to a global demand for mutual monitoring, simply because this is the only thing that is left to do on Earth. Just consider the time and energy that can be spent on Twitter to comment on the slightest changes of any star's hairstyle. Watching the life of others have become one of the most important activities on Earth. While we're watching, big data companies are recording and digitizing just about every piece of our lives, right down to our intimate behavior and facial features. In the near future, our slightest gestures will be indexed, analyzed and then anticipated. As the key source of income for new tech companies will come from the capacity of their algorithms to predict and thus orient our purchases, any efforts to escape these surveillance zones or deviate from expected consumer behavior will be considered nonconformity and rebellion.

There is actually nothing to suggest that today's technological innovation will be put to good use in the future. There is no science-fiction movie that anticipates a flourishingly positive usage of technology. Instead, despite its many benefits, in healthcare for example, it is presented as a two-headed monster: generalized control over humans on the one side and out-of-control artificial intelligence on the other.[12] At the current pace of technological progress, our subjectivity will soon be the only thing left to distinguish us from machines capable of doing almost all of our essential activity, right up to giving orders to workers under their command, whether it be on factory floors or in business district offices.

As always, machines bring the good with the bad, depending on how they are used and the extent to which human beings will submit to artificial intelligence.[13] The daily spectacle of workers glued to their computer screens and smartphones

---

11 No wonder that the panopticon was invented by utilitarian thinker Jeremy Bentham, whom I already portrayed as the precursor of modern capitalist brutality. The panopticon is a type of building that allows a single watchman to observe the inmates without them being able to tell whether or not they are being watched. Big data companies are doing just that when waving their so-called privacy policies to let people think they're not being tracked.

12 On an enclosed planet, the extension of technology's reach right into the most intimate aspects of our lives and souls will not make us any happier. Take the mobile apps that track users in supermarkets via their smartphones in order to send them commercials in real time: are they really making our world a better place?

13 IBM's information system assistant for decision-making, called Watson, already gives us an idea of the capacities developed in this field.

does not bode well for the future. It will, I am sure, also depend on the objectives we assign to these machines: if, in our closed world, it is a case of surpassing our intelligence and creativity and improving the way they survey and monitor our every thought, as a lot of major big data companies already do,[14] then it is bound to end in tears. We will thus be changed, not into fairy-tale frogs but into binary codes and quantic flows, immortal but with no soul, or just enslaved by tyrannical algorithms. If, however, they are designed to take us spaceward, enabling us to send interstellar probes and robots, to calculate trajectories, manage ultra-complex propulsion systems and transmit images of faraway planets, then the technologies that enslave us could well again become our most precious instruments for exploration, and humanity will have the opportunity to take the power back.[15] If we wish to remain free in the twenty-second century, we will have no other choice. If we do not, progress will consume us.

Whatever happens, the only escape from the existentially barren world currently portrayed in cinema fiction appears to be the route back to square one, whereby a small group of men and women, having slayed the serpent, are given the task of bringing Genesis back to life. Is reinventing the world not the hidden dream of humankind? Every child delights in inventing stories from nothing, calling upon our creative skills and sense of kinship, where we triumph through courage and resourcefulness, like the Hardy Boys or Tom Sawyer. But for children born today, what hope is there for them to create their own destiny? I am not nostalgic for yesteryear, but I am aware that there are worlds, landscapes, moods and feelings that the world has left behind, the likes of which will never be seen again. This is the price of human progress and a repeated catastrophe for humanity: we can feel some forms of happiness slowly waning,

---

14 Those companies already rule our world. Denmark announced in January 2017 that it would name a "digital ambassador" to deal with giant technology companies like Facebook and Google. Those companies "have become a type of new nations" and "affect Denmark just as much as entire countries," Foreign Minister Anders Samuelsen said in an interview with Danish newspaper *Politiken* (January 26, 2017).

15 The big bosses of digital industries are the first to ensure their children are kept well away from screens and tablets, sending them to schools with good old chalk and blackboards. These people, the masters of today's world, have understood a vital point: that passivity is the driver behind the digital society. To make people accept this passivity, one has to make them feel proud of their behavior and hence persuade them they are "in," "cool," "in tune with the times" and with progress in general. The countries who will dominate the world of tomorrow will be those who have understood just how to protect their population from widespread technological brainwashing and who invest in the fundamentals of education that have been with us since Antiquity.

one of them being our capacity to forge a bond between our bodies, minds and nature itself. We know we cannot go back in time, and the near future is looking less than appealing. I often wondered why the novel took so long to appear in human history. I now know why: the novel was invented when humanity realized the importance of immortalizing what would never be experienced again.

To escape this enclosed, predictable, unfair, polluted, violent and controlled world, humanity's greatest fantasy remains the flood myth. A way to destroy everything and start over, reemerging from our caves to relight the fires of desire, knowledge and the search for utopia. The real hero of humankind is Noah. It was remarkable to see that two movies were released in the same year, 2014: first, the eponymous story depicting the biblical narrative of the Great Flood,[16] and second, *Interstellar*, which portrays humanity's departure from Earth to another world. These scenarios describe the two options left to us to experience a new world, while at the same time expunging all our ghastly sins. There are now an endless number of movies that depict societies devastated by nuclear wars, pandemic viruses or environmental destruction, leaving humanity contaminated, irradiated and mutated, either destined to kick-start the species from scratch or obliged to leave Earth. The real subject of these stories is not the fear of a possible extinction of the human race, but rather our secret hope that only through the most complete of adversities will we be obliged to rebuild a new world that has learned its lessons from the past and has recreated the right conditions for a new collective adventure, avid for redemption. Just as in Genesis, when God sent the Great Flood to give a second chance to His flawed and heartless human creatures. Yet we have never seen such a flourishing number of survivalist movements, heralding humankind's arrival in a state of permanent millenarianism.

16 *Noah*, directed by Darren Aronofsky, 2014.

PART III

THE GREAT FILTER

CHAPTER 7

OVERCOMING DIVISION

Our world is ending, and humans are bored. Sat in their offices, their desire has withered away. To forget their depression, they become frenzied consumers, but this comforts them less and less. A time will come when they will throw themselves into a rage, hammer on the walls, strike down others in their fury, go crazy and dream of revolution, the apocalypse and the Great Recommencement. Salvation may lie in our departure, but leaving Earth is easier said than done. When reading such a sinister description of our future, one may wonder if there is any hope at all. The answer is yes, there is, although huge obstacles stand in the way of our expansion into space.

Optimists could argue that technological progress has accelerated so much in the recent past that we should not worry too much about our ability to reach for another star. They could also say that we went to the moon in 1969 with technology that was far inferior to mobile phones we were using in 2010 and that mobile phones themselves did not even exist when we sent *Voyager I* to space in 1977. Based on this trajectory, space travel would get cheaper and cheaper and increasingly efficient.

I do not fully agree with that argument, for three reasons. First, I believe that interstellar travel confronts us with a technological gap that has nothing in common even with the one that separates the wheel from nuclear propulsion. Moving at the same speed as *Voyager I* (at around 60,000 km per hour, 17 km per second or 0.005 percent of light speed), reputed to be the fastest machine ever created by man, and the first to have reached interstellar space, it would take 75,000 years to reach Proxima Centauri, the closest star to our solar system, located 4.24 light-years from Earth.[1] If we were to attain 10 percent of light

1 Scientifically speaking, it would be more accurate to point out that it would take 75,000 years to cover the distance between the sun and Proxima Centauri. For a spacecraft to be able to reach Proxima Centauri, one would have to take into account acceleration and

speed, we could in theory hope to reach this coveted star in 42 years, from a terrestrial viewpoint, which would make it a one-way trip.[2] To bring a person alive to a point situated light-years away from our planet and return will require to master sources of energies that are unknown to us today, and this will imply a gigantic technological leap forward.[3]

Second, the cost of the technologies needed for space travel will be so high that the amount of money involved would by far exceed what has been spent in the past for any kind of exploration. It is impossible that one country alone would succeed in gathering the financial and technological means necessary to create an interstellar space program, regardless of how powerful they are. Let us take an example. According to Gilles Dowek,[4] the fabrication of integrated circuits to be found in our computers represents billions of dollars, compared with "only" millions of dollars for car manufacturing. The cost of the most miniaturized circuits is so high that they are produced only in a few factories around the world, made from plans drawn up by other companies.[5] In short, "building a factory for integrated circuits goes beyond the investment capacities of any individual person, family or even country: it would require several million people investing together, which would need a certain coordination."[6] This coordination is done today through financial markets. It is clear that an interstellar space program would, among other challenges, require the production of miniaturized circuit boards on a level that no state would be able to do alone, not forgetting the cost of related research into propulsion systems.

One of the methods proposed to shortcut the problem of insufficient public resources is the privatization of space exploration. Elon Musk, CEO

---

deceleration times, on the one hand, and the trajectory of Proxima, on the other hand. Like other stars, its distance from the sun is constantly varying. Proxima Centauri may today be our closest star, but this has not always been the case. According to papers published on February 17, 2015, in the *Astrophysical Journal Letters*, a red dwarf by the name of Scholz's Star (to honor the man who discovered it) came within a distance of 0.8 light-years from our sun around 70,000 years ago.

2 Given the effects of special relativity, the time elapsed for the astronauts would be less.

3 To paraphrase Neil Armstrong's famous statement on his arrival on the Moon on July 20, 1969: "That's one small step for a man, one giant leap for mankind."

4 Researcher at the French Institute for Research in Computer Science and Automation (INRIA) and member of the scientific council of the French Society of Information Technology.

5 There are six in the United States, four in Taiwan and one in France, Germany, South Korea and Israel.

6 Gilles Dowek, "Homo habilis hypercapitalisticus," *Pour la Science*, no. 446 (December 2014).

of SpaceX, considers that the implementation of a real space economy will be the only way to successfully finance the colonization of Mars, including "terraforming" (in his mind, we would need to detonate nuclear bombs there in order to release enough gas and water from the Martian poles to warm its atmosphere), something he judges indispensable to save humanity from its inevitable extinction. Commercializing space exploration is more than feasible, whether in the form of space tourism, telecommunications development or the exploitation of new resources (for instance, in February 2016, Luxembourg announced that it wished to engage in exploiting mineral resources from asteroids). But it is unthinkable that a project such as interstellar travel could ever be fully funded through only private means, even when imagining a business model on a scale of our solar system. Business models for American space companies, massively financed by NASA, are more about externalization than privatization. If lucrative or recreational space projects were to become prevalent, this would lead to a drop in any research and exploration capacities, which couldn't provide commercial companies with a rapid return on their investment. But humanity's reason for traveling to another star will not be money. A voyage beyond our solar system will involve such a huge investment that, as for most fundamental research projects, only a powerful consensus of public investors and big multinational companies will be able to generate the finances necessary to come up with the required technology. The question of financing a manned mission to Mars, announced by NASA for the 2030s, gives us a good idea of the amplitude of the problem involved, as it could cost up to 1,500 billion dollars according to the most pessimistic estimates. The United States will not be able to come up with such money on their own, which would therefore involve international cooperation on an unprecedented scale.

Third, the budget for space exploration will be so high that it is unlikely to be easily approved by voters, as it was rightly foreseen in *Interstellar*, where the whole space program has been hidden away and propaganda activated to make people believe that walking on the moon was a myth. The balance between money spent on space exploration and on fighting poverty or promoting education is already difficult to find, and this will be more so in the future. But if we had not found the means to allocate the necessary resources for space exploration in the past, we would have never set foot on the moon. The world would probably be poorer, having failed to invest resources in innovative technological cycles, which require injecting large amounts of money into projects that bring no short-term returns and have no immediate application.

Wandering through space will therefore be much more than a giant leap forward. So if we really wish to surmount the technological and financial limitations preventing us from traveling to the stars, humankind will have to confront, and defeat, one of its oldest demons and, in my view, one of the major components of the Great Filter: division. I have argued earlier that a global polity could not be anything else but a global dictatorship, and this is generally what anticipation movies or novels show. But can humankind go to space as a species without uniting in a global polity? One may of course argue that, if traveling to space really is the only way to escape death on Earth, this objective could justify global dictatorship (you may call it an empire to make it more presentable, but its founding principle of getting rid of democratic representation would remain intact). To avoid this daunting dilemma, I would like to defend the idea that division can be overcome without necessarily forming a global authoritarian polity and that we can have world unification without global domination.

I beg the reader to forgive me to take examples not from prominent political thinkers, but from a Hollywood blockbuster movie, *Independence Day: Resurgence*,[7] in which the aliens come back to take their revenge after having been defeated by the humans 20 years earlier. As a result of their initial defeat, humans have decided to end their wars and cooperate on an unprecedented scale to build a common global space defense against future alien attacks. In the movie, the United States takes the lead of this new organization, but other major nations, especially China, still have a major say. The basic political units of this future world are still nations, not a global polity. This might be quite a simplistic demonstration compared to the sophisticated analyses of political philosophers, but it tells us something important: that even when faced with the deadliest dangers, even our imagination is not ready to surrender political freedom and sovereignty, not only because we cherish them (not everyone, I admit) but because we feel that their demise would destroy us more surely than any alien army.

Why am I insisting so much on warning against the global polity utopia? One could argue that we could just try it and see if it works. But trying to make it work by all means is exactly the way to undermine the beautiful ideal of unification for a long time. The British exit from the European Union has proved that forcing political integration leads inevitably to disintegration. This is what happened to all the great empires of the past, and this is what could happen to the European Union, starting with the dismantlement of the Euro zone, as Nobel Prize–winning economist Joseph Stiglitz has warned.[8] As for

7 Directed by Roland Emmerich, 2016.

8 Joseph Stiglitz, *The Euro: How a Common Currency Threatens the Future of Europe* (New York: W. W. Norton, 2016).

all utopias, the road to hell is paved with good intentions. The unification of humankind may be a good objective in theory, but unification, understood as forcing everyone to obey the same global rules, may well become the worst totalitarian nightmare in human history. Other types of totalitarianism could anyway emerge, as I suggested earlier, for instance, in the form of a global surveillance society, concealed behind the benevolent traits of consumer welfare.[9]

The conquest of space has always been a bone of contention for the world's superpowers. But it is also one human project that has already helped build the sturdiest of collaborations—one example is the Apollo–Soyuz mission, which, on July 17, 1975, saw the first handshake in space between an American astronaut (Thomas Stafford) and a Russian cosmonaut (Alexei Leonov). Whereas space programs were for long exclusively national, many are today multinational, especially in the research field. Astronauts from various nationalities work together in the International Space Station, and they were brought up there by various spacecrafts (US space shuttles, Russian Soyuz spacecrafts and since 2020 Space X Dragon 2 spacecraft). The European space agency (ESA) comprises 22 members. China and ESA are discussing potential collaboration on a lunar base.[10] NASA and its Russian counterpart Roscosmos signed in September 2017 a joint statement for "deep space exploration" supporting research that could lead to a cislunar habitat. This need for cooperation has become so obvious that it has even been integrated by science fiction films, for instance, in *Arrival*[11] or *The Martian*,[12] in which NASA has to seek help from the Chinese to save their astronaut stuck on Mars. This reminds us about one thing that interstellar dreamers often forget: space is money, a lot of money. Crowdfunding in sight.

If we really hope to one day leave our solar system, we will need more than mere handshakes. There will always be divisions in the future, and nationalistic behaviors will endure. But cooperation can progressively round out the angles.

---

9 The Chinese state is setting up a vast ranking system that will monitor the behavior of its population and rank them all according to their "social credit" in association with rewards and punishments (banning you from flying or taking the train, banning your kids from the best schools, throttling your Internet speeds, etc.).

10 Matthew Brown, "China Talking with European Space Agency about Moon Outpost," *Bloomberg*, April 26, 2017.

11 Directed by Denis Villeneuve, 2016.

12 Directed by Ridley Scott, 2015. The movie was based on a novel of the same name by Andy Weir (New York: Broadway Books, 2014).

This will not depend on any global bureaucracy, but on the wisdom of the world leaders in the next hundred years. They will have the power to finish the ecological destruction at work and start the Third World War, or to bring humanity to its next evolution step, making sure that scientific progress does not turn us into submissive and dehumanized mutants.

CHAPTER 8

FIGHTING RELIGIOUS ANGST

Overcoming division will be the easiest part. When technical progress brings us closer to a possible departure, I think the other major obstacle will actually be the considerable religious angst that will surely arise. Even though major religions aim at building "palaces in time,"[1] they are all of this Earth, born in specific places with specific roots. In Ancient Greece, the Gods lived up on Mount Olympus. In Genesis, heaven is located in Eden, Mesopotamia. Moses lived between Egypt and the Promised Land, Jesus in Palestine, Mohammed in Arabia and Siddhārtha Gautama in Nepal. Can we take God, or rather our gods, up into space? If we settle on another Earth, will we build churches, synagogues, temples and mosques? Will the men and women who colonize other worlds come back to Earth as a pilgrimage? Will they look toward the Blue Planet as they say their prayers? No major religion appears to clearly mention any other worlds, inhabited or not,[2] but there is nothing that would a priori oppose the concept of uprooting them and transferring them to another planet.[3] At least not in theory, for in the majority of humankind's inner faith, at least as

---

1 I have taken this expression from Abraham Heschel, *The Sabbath* (New York: Farrar, Straus and Giroux, [1951] 2005), for whom the essence of Judaism is to build a "palace in time" thanks to the Shabbat, whose spiritual preeminence is taken from the order by which the Creation was sanctified by God in Genesis: first time, then man, then space.

2 The Catholic Church admits the possibility of the existence of extraterrestrials, but the question remains open as to their status within Creation, not unlike the way Neanderthal man and other hominids are perceived. Do they have a soul, the church wonders? Are they concerned by original sin? These questions may have seemed theologically preposterous in another age, but since we have discovered exoplanets and know we have inherited certain Neanderthal genes, this is, from a religious perspective, no longer the case.

3 For example, the majority of Christians keep their faith without feeling the need to visit the Church of the Holy Sepulcher.

I perceived it when growing up, man is the only subject of Creation and Earth his only home.

Indeed, the Catholic education I received was a perfect reflection of how science had described the Big Bang, this monumental fiat lux that had provided the prologue to Creation itself. Without knowing it at the time, I was a concordist, in the same way that Pope Pius XII had viewed the burgeoning theory of the Big Bang, as initially developed by Abbot Georges Lemaître in the 1930s, as a scientific confirmation of the story portrayed in Genesis.[4] In every scientific discovery, I only ever saw reaffirmed proof of the existence of God and, furthermore, His project to make humankind the ultimate and exclusive goal within the universe. For instance, I perceived the extinction of the dinosaurs as a guided quirk of fate, because had the killer meteorite been just a little bigger, smaller life-forms would have been wiped out along with the bigger reptiles

---

4 In a speech given on November 22, 1951, at the Pontifical Academy of Sciences, amateur astronomer Pope Pius XII spoke about the Big Bang Theory (without actually naming it), declaring: "In fact, it would seem that present-day science, with one sweeping step back across millions of centuries, has succeeded in bearing witness to that primordial 'Fiat lux' uttered at the moment when, along with matter, there burst forth from nothing a sea of light and radiation, while the particles of chemical elements split and formed into millions of galaxies." In his speech of September 7, 1952, at the World Astronomy Congress, he did not reaffirm this position, but instead preferred to insist on the absolute otherness of the Holy Spirit, which observation of the cosmos enables humankind, in his view, to discover. At the same time, he also paid tribute to modern science, without fear of being critical, somewhat abrasively, of the Catholic Church's former attitude toward major cosmological discoveries:

> As for us and our solar system, we are not at the center of this immeasurable scattering of stars, as was previously thought: we are in fact 30,000 light years away. [...] As a result, the amplitude of this cosmic realization, which legitimately deposes the geocentric and anthropocentric ideas of old, reducing our planet to the size of a grain of astral dust and mankind to a mere atom, and relegating both to some lost corner of the universe, is not an obstacle to love (as some have affirmed when discussing the mystery of the Incarnation) or to the almighty power of He who, being pure in spirit, possesses an infinite superiority over matter, regardless of the cosmic dimensions in space, time, mass or energy. [...] May the modern conception of astronomical sciences, which has represented the very ideals of so many great men throughout history, from Copernicus to Galileo and Kepler to Newton, continue to flourish and provide wonderful progress to modern astrophysics.

On October 31, 1992, Pope John Paul II recognized the church's mistakes concerning Galileo, but in 1999 refused to reinstate Copernicus.

and mammals would not have been able to evolve and give birth to human beings—like Goldilocks's final chair, it was just the right size to enable them to rise up from the Jurassic swamp. Finally, fed by biblical stories and classical culture devoid of any reference to other worlds whatsoever, I did not believe in aliens, and I could not for one second imagine that we would one day be able or even wish to leave Earth. Of course, I was curious to learn about all kinds of scientific discoveries, but the space race didn't interest me in the slightest. For me, mankind's giant leap to the moon was nothing more than a small technical exploit of slightly megalomaniac proportions, with neither utility nor particular relevance to the future of humanity.

In fact, I felt protected within my terrestrial cocoon, as if the whole universe's only raison d'être were to contain our planet and provide us with shelter. Earth seemed to me to have only one unique role, this theatrical stage allowing humankind to play out history's greatest drama, under the watchful eye of a God who would arbitrarily decide just when to bring down the curtain—in this case my mind would bring down its own chaste veil upon the implications resulting from such a postulate, preventing my imagination from roaming a thousand, thirty thousand or a million years into the future. The slow emergence of stars, planets and then the evolution of life on Earth appeared to me a necessary chore accomplished by a conscientious and very loving God who, by letting matters run laboriously until the arrival of humankind, wished to prove to us that we were indeed free. A lazy and tyrannical God would have created the world in six days, without any consideration for the delicate minds of these poor humans, crushed down by the weight of such power.

As my teenage years came to an end, my childish cosmogony veered away from those soothing representations to find root in philosophical thought, while still guided by my Christian upbringing. This was in fact proof that cracks were silently beginning to appear in the foundations of this impressive edifice, fissures that were calling for help, some sort of consolidation or intellectual justification. For these unvoiced doubts, which had been plaguing my subconscious for years, were discretely beginning to take shape. I started to imagine Earth seen from space instead of from a Western perspective or one based on Revelations, and I became aware of a human adventure embedded within a much vaster history. What I had learned at Bible lessons no longer seemed to be capable of confronting the mystery of the human condition within the universe. I didn't get as far as reaching the conclusion that what I had learned was false, but rather that God, having decided that we were free, had chosen to give us the possibility of doubt and the understanding to confront it. So I threw myself headlong into developing passionate ideas based on philosophical concordism, spending hours reading philosophical works and making pages of scribbled annotations.[5]

Despite all my efforts to protect my childhood beliefs, rents began to show. The first, completely unannounced, appeared while on a visit to Scotland. I was 23 years old, studying in London for a year, and had decided to make the most of the school holidays to spend a few days' trout fishing up near Perth. As I was walking alone under a dark gray sky along a river lined with gnarled trees, a little lost in my thoughts, I was suddenly overcome by a rush of dark, black emptiness, brought on by the harsh realization of just what the immensity of the universe really meant. The infinity of space-time had brutally switched from an intriguing philosophical concept to a physical and spiritual sensation of my complete and utter perdition in the vast interstellar void. Faced with the impossibility of being able to cling on to anything that could hold the universe, I was overwhelmed by a powerful feeling of vertigo, as if the earth itself were going to collapse in a horrifying cosmic calamity. I was gripped by a blood-chilling feeling of utter dread, which still returns whenever I think back to this moment. My faith was permanently shaken by this. Such a shattering feeling of the total absence both of God and of myself, after years of intellectual and emotional certainty of His presence and love, as I perceived Him as a person, attentive to my voice, was like discovering my mother and father weren't my real parents. It was like a kind of mystical shock in reverse, for which nothing had prepared me. Some years later, I accessed the rational description of the immensity of the universe allowed by astronomical instruments and discovered that the only thing that enables us to place any kind of limit on this vastness is our observation capacity—and, disturbingly, we cannot see any.

The second fissure became apparent when I realized that the sun, and therefore Earth, would one day come to an end. The sun's ultimate combustion will, of course, not take place for several billion years, but Earth will become uninhabitable well before then, in around one billion years' time, when all its water will have evaporated and the surface temperatures will have become

5 Without knowing it, I was appropriating the stance of Dominican friar and theologian Yves Congar (1904–1995), who denounced the "suspicion towards intelligence" emanating from a church whose own intellectual life in the Middle Ages, thanks to the disputations, seemed paradoxically to be much richer when it held unwavering domination over the minds of men. Yves Congar said, "we are not accused of having written something in particular, but simply to have written anything at all" (cited in Bernard Chassé and Laurent Lapierre, *Marcel Brisebois ou le musée d'art contemporain de Montréal 1985–2004* (Québec: Presses de l'Université du Québec, 2011), 44).

unbearable. This eventuality may seem far-off, but this is very soon when seen on a cosmic timescale, and it means that life on Earth, having first appeared 3 billion years ago, has already lived through the majority of its existence and is moving toward its death throes. The demise of Earth confronts us with something far beyond our own mortality—the end of humankind itself. We humans now know ourselves mortal.[6] Unlike with a natural death for Earth, I took great reassurance in the end of the world according to Christian beliefs, or the Apocalypse, as it demonstrated the most coherent attribute of God's almighty power: that of being able to take back what He had given. Furthermore, this ultimate and scripted intervention led me to the false belief that nothing dramatic could happen to humanity until this came to pass. However, upon learning that we humans were destined to die along with the sun, I was confronted with a formidable dilemma: maybe God had programmed the end of the world before our star imploded, which would leave us little time to live. Either this, or maybe God had decided the world would die from natural causes, which abrogated the idea of the Apocalypse and for me opened up an abyss of doubt regarding His project and divine love: would our Lord really have created the whole universe simply to allow humans to exist laboriously for a few hundred million years at best, before letting them die out in huge suffering, leaving behind an icy void for the rest of eternity?

The third upheaval appeared with the discovery of exoplanets. In the 1980s, astronomers had not yet discovered any exoplanets at all, and some even doubted that they could exist. It therefore seemed natural to me to think that Earth was the only inhabitable planet in the universe, and surely the only place where life had developed. The Fermi paradox only confirmed this hypothesis. I remember the scientific community as being largely convinced of the uniqueness of our solar system's configuration, not to mention that of Earth, located conveniently within an orbital zone providing the best conditions for life to thrive.[7] In 1995,

---

6 To paraphrase Paul Valéry, who wrote in 1919: "Nous autres, civilisations, nous savons maintenant que nous sommes mortelles" [we civilizations now know ourselves mortal] [*Variété I* (Paris: Folio, 1998)]. At that time, in the aftermath of the First Word War, his statement applied only to the big cultural units of humanity that we call civilizations.

7 On the scale of our solar system, our planet is indeed not only situated within our sun's habitable zone; it also benefits from other extraordinary characteristics that have probably proved decisive in the development of intelligent life-forms: here on Earth, the presence of sufficient liquid water for life to develop but not enough to prevent continents from emerging (had the planet stayed covered with oceans, it is not sure humans would have appeared), a magnetic field (the Van Allen belt) that is strong enough to protect us from solar winds and a sun–moon combination creating the tidal movement that is said to have contributed to

however, the first exoplanet was discovered, and since then, nearly 3,300 others have been officially listed,[8] and this with such accuracy that we are able to determine an approximate makeup (gaseous or telluric) and the distance away from their star, thus in turn a supposed degree of habitability.

This abundance of exoplanets shattered my geocentric view of the universe as, all of a sudden, I became aware of the possibility not only that life could have developed elsewhere in space but also that man could live on another Earth. This realization shook me to the core. I could no longer refuse to believe in the existence of other planets or of any form of extraterrestrial life on the grounds that they hadn't been discovered. I would have to do so purely because it appeared incompatible with my vision of a divine project rooted in an exclusive relationship with terrestrial man that set us at the heart of Creation itself. I viewed the hypothesis of another "humanity" as a spiritual monstrosity. If life could rise up or flourish elsewhere than on Earth, then our planet would lose its status as humankind's divine vehicle and become a mere natural phenomenon, of which we would be nothing more than its most evolved life-form.[9]

---

the development of life and to ensure our planet maintains its rotational axis, thus allowing for seasons and an overall temperate climate to exist. As far as the sun is concerned, its formation appears to be due to the unusual conjunction of various cosmic events (a nearby supernova, a massive star, a gigantic stellar cluster and a Wolf-Rayet star), its radiation is highly stable and provides solar winds, which protect Earth from cosmic rays. And finally, our whole planetary system is very protective, notably thanks to Jupiter, whose force of attraction keeps the asteroid belt continually in its gravitational field, thus preventing Earth from constant bombardment. Obviously, we are a statistical miracle, unless the universe is infinite or there are billions of universes both in space and in time or we find countless intelligent life forms dwelling on numerous exoplanets.

8 In May 2016, NASA announced that 1,284 new exoplanets should be added to the 2,000 already discovered (and many more are just waiting for official confirmation). According to data from the Kepler Space Telescope, 22 percent of the 150,000 observed stars possess a planet the size of Earth, located at a distance permitting the presence of liquid water. This has led certain scientists, somewhat hastily, to point out that, by extrapolation, there could be 9 billion potentially habitable planets in the Milky Way. Let us not get carried away: for the moment, we know absolutely nothing.

9 I was convinced that science could never explain the emergence of life—and actually it still cannot. The truth is, I was scared by the idea that life could appear on its own, without divine intervention, right from the star dust we're made of. If such a discovery were made, it would not bring an end to the quest for the origin of life but only shift it to the inner parameters of matter itself, without necessarily bringing to light the missing link between physics and biology.

The last source of questioning emerged when I realized that we humans of the *Homo sapiens* species were not descendants of those prehistoric silhouettes progressively straightening themselves on the evolution poster that was pinned to the classroom wall, but the surviving siblings of the Homo family. In other words, we were not alone: there was a time when there were several kinds of humans. As Yuval Noah Harari writes,

> It's a common fallacy to envision these species as arranged in a straight line of descent, with Ergaster begetting Erectus, Erectus begetting the Neanderthals, and the Neanderthals evolving into us. This linear model gives the mistaken impression that at any particular moment only one type of human inhabited the earth, and that earlier species were merely older models of ourselves. The truth is that from about 2 million years ago until around 10,000 years ago, the world was home, at one and the same time, to several human species.[10]

So, in the end, who was Adam? If I had always welcomed the idea that man was part of the natural history of Earth and was also submitted, like all living beings, to the evolutionary process, I was not ready to admit that God had to make several drafts of us, all endowed with a sort of conscience and soul, before allowing glorious sapiens to become the chosen one, or that we were just the winning result of sheer animal competition. The spiritual monstrosity I feared to find in space (the possibility of other humanlike beings living in another planet) had actually already occurred here on Earth![11]

And so it was that an entirely new vision of the world progressively presented itself to me, moving away from that of a permanent and steady universe (to use a word from the eponymous theory)[12] harboring an eternal planet serving as

---

10 Noah Harari, *Sapiens*, 8.

11 On October 15, 2015, a NASA study announced they had observed irregular light variations from a star located 1,420 light-years from the sun, which could suggest, among other hypotheses, the existence of an artificial megastructure. But no radio signals from this star have been detected since. For the moment, I feel reticent regarding the idea that the universe is inhabited by other beings equipped with a conscience. Maybe this is another of my "childhood truths" that I do not wish to relinquish.

12 The Steady State Theory, expounded in the 1940s by Fred Hoyle, Thomas Gold and Hermann Bondi, postulates that the universe is identical at any time as well as at any place, in other words ageless (or eternal) and unchanging. It was gradually left by the wayside as the Big Bang Theory grew in popularity, due to the fact that it was incapable of taking into account, despite adapting to the theory of a quasi-static universe, observations concerning the fact that far-off galaxies were thought to be moving away (see the following note).

a garden for mankind, to whom history belonged, whose only goal was more or less to master our destructive impulses in order to put an end to the endless succession of wars going all the way back to the symbolic clash between Cain and Abel. I was beginning to realize that the universe too had a long history, one which went much further back than ours and was continuing in front of our very eyes: stars are being born while others are dying, galaxies are diverging and merging, the frontiers of the universe are perhaps still on the move and space-time itself appears to be growing.[13] Earth's very history has been marked by terrible cosmic cataclysms, such as the probable collision with a planet the size of Mars, which is thought to have created the moon. Our solar system itself is moving across a galaxy that is drifting across a universe that is apparently expanding, and we can even wonder if this universe is not in turn traveling across other universes. Albert Einstein himself was extremely reluctant to abandon his belief in the existence of an immobile and eternal universe. When designing the general relativity theory, he introduced a "cosmological constant" into his equations in order to neutralize the implacable results before his eyes, that of a universe which had at one point been very dense and hot, meaning it therefore had both a beginning and a history. It was to be the "greatest mistake of his life," as he himself admitted following the discovery of the cosmic microwave background in 1964.[14]

In fact, science teaches us that everything in the universe is born, undergoes transformation and then dies. This is the way that science has overturned ideas from the ancient world, which was dominated by ideas of permanence, fixity, eternity, immobility and regularity—as cataclysms were interpreted as anomalies, inexplicable incidents or unnatural upsets, it was thought that they must have been brought on by God's divine wrath. Our ancestors perceived movement and change as swinging, recurrent and circular, not unlike some giant clock controlling the universe. It is hardly surprising that the eighteenth century, benefiting from the rediscovered Antique world during the Renaissance, applied

---

13 The expanding universe theory is the most convincing hypothesis to explain, within the theoretical framework of general relativity, the way that the furthest galaxies (those located beyond the gravitational attraction field of the Laniakea Supercluster, which is 500 million light-years wide) are shifting toward the red of the light spectrum.

14 This concerns the residue from the very low-temperature electromagnetic radiation theoretically left over from the Big Bang, the existence of which was predicted in the late 1940s by Ralph Alpher, Robert Herman and George Gamow, and then confirmed through observation by two engineers from Bell Laboratories in 1964 and later on by the COBE, WMAP and PLANCK satellites.

the notion of the Great Watchmaker or the Grand Architect of the Universe, whereby the watch's adjustments were portrayed as much more important than its fabrication, something which back in those days no one imagined could take up so much time.

Thanks to scientific progress, we are slowly becoming aware of the precariousness of our astral island, whose physiognomy we now perceive much like some fragile organism that is born, then grows and will then one day perish, after having undergone ordeals, diseases and a few other bumps and bruises along the way. Humanity has lived long enough and has the sufficiently powerful measuring tools required to know that even its terrestrial environment is in constant change: the magnetic pole is shifting and weakening, the climate is changing, the continents are drifting,[15] the atmosphere is leaking away,[16] our moon is moving further away,[17] the length of the day is increasing,[18] our sun is aging and species are constantly appearing and disappearing. What is also harrowing is the observation that the majority of living species appears to have a life expectancy: if this is so, why would the human race stop evolving? Why would it remain fixed in a biological state that no other species has ever been able to conserve, apart from the most primitive? Even though, in my view, our fixed

15 According to estimations from the Paleomap Project (University of Texas), in 50 million years' time, Africa will have come into contact with Europe, closing up the Mediterranean Sea. The Red Sea will have been swallowed up by the Arabian Peninsula, which will also be joined to Africa, whereas the widening of the Atlantic zone will have pushed the Americas further across the Pacific.

16 The current depletion is around 3 kilograms of hydrogen and 50 grams of helium per second. These amounts may be tiny, but by no means insignificant when seen on a cosmic scale, as shown by the massive evaporation of the atmosphere on Mars during its early years which, according to the MAVEN Mission (Mars Atmosphere and Volatile Evolution Mission) made public on November 5, 2015, was brought on by solar flares.

17 This is a rather unknown phenomenon, but a lot easier to define than the aging of the sun. Laser reflectors were set up by the 1969 Apollo 11 mission, providing us with the estimation that the moon is moving away from Earth by 3.8 cm per year. Earth's deceleration, created by tidal friction (due to the moon and the sun), would cause the moon to slow down proportionately, thus extending its orbit.

18 The mean length of a day is increasing by about 1.7 milliseconds per century, due to the tidal acceleration effect on the moon. The length of a day circa 1 billion years ago has been estimated as having been about 18 hours and 41 minutes. This phenomenon should stop in a few million years, when the moon will be visible only by half of the planet (Stephen R. Meyers and Alberto Malinverno, "Proterozoic Milankovitch Cycles and the History of the Solar System," *Proceedings of the National Academy of Sciences* 115, no. 25 (June 2018), 6363–68.

and millennial vision of an eternity safe on solid ground remains predominant in the collective psyche, it is slowly shifting to that of a vast sailing boat, one that is certainly well-built, but easily tossed about by the elements and not able to prevent being capsized. If early humans' fears came from the unknown, from all those things they could not explain, then ours will more likely come from what we can now foresee, revealing the vulnerability of our terrestrial raft as it drifts across a universe as turbulent as a wild, unforgiving ocean. Thanks to the immense progresses of science, to reflect upon the human condition today is no longer a case of admiring the starry skies from Earth, but of observing Earth from the deepest, darkest space.

To sum up, I believed as a child that we humans were unique, that Earth was our safe and eternal home and that God cared for our destiny (well, He killed the dinosaurs). And I was finally brought to admit that we might not be that special, that Earth was more of a rowboat than a harbor and that all God's love would probably not deviate asteroids nor help us build interstellar spaceships to fly away in case of danger.[19] Of course, our brains already knew that, but have we spiritually accepted the fact that Earth is doomed and that we may just be one of the million life-forms in the universe? Have we really confronted our deepest beliefs with what science tells us about our situation in the universe, the origin of man and the destiny of our solar system?

---

19 Actually, the anthropic principle obliged me to add a new episode to my spiritual quest. This is the hypothesis that the universe's parameters (such as its density, respective masses of neutron and proton, balances between gravitational, electromagnetic, weak and strong nuclear forces) are set with such a high degree of precision that if any of them were to undergo even the tiniest of modifications, then the universe would be incapable of sustaining life in general, and man (*anthropos*) in particular. For example, had the universe been denser, it would have collapsed in on itself, and had it been less dense, the force of expansion would have led to the disintegration of every stellar or galactic body. In other words, this theory, in its strongest version, states that the universe has been tailor-made for man. Critics reply that this is the very definition of a tautology. But this theory pushed me to consider the possibility that we could eventually be quite special and that the universe could actually look like Goldilocks's final chair. This leads many people to believe that God, just 200 years after Laplace's demon, is in fact a scientific hypothesis among others, for only two possibilities remain: either an omnipotent creator has set the mechanism running, or our universe is simply the winning lottery ticket among millions of others. The theory of multiverses, stretching over both space and time, is closely linked to the will of those who seek to return "chance" to its former status of philosophical preeminence in the field of science. In any case, never has the alternative between chance and necessity been presented with such clarity and contrast.

In 1755, a terrible earthquake took place in Portugal, which almost destroyed the city of Lisbon. Between 30,000 and 40,000 people were killed and many precious documents and buildings were lost forever. This event caused tremendous emotion among the populations, for it stroke the heart of Roman Catholic Europe on the morning of November 1, the holy day of All Saints' Day, causing speculations on its religious meaning. This catastrophe was seen by the philosophers of the Age of Enlightenment as evidence that the benevolent God of the Christian tradition should give way to the figure of a great Architect who left nature work on its own and endowed us with reason to understand how. Even if it took a lot of time, European peoples' faith was indeed deeply transformed, and among today's European Christians, you wouldn't find many seriously arguing that the 2004 Indian Ocean Tsunami, which caused the death of nearly 230,000 people, could in any way be attributed to a divine vengeance.

Christians in the eighteenth century had no other choice but to confront what they believed or thought to know about God with that natural catastrophe and therefore admit that there was a limit to God's scope of intervention in human history. In the same way, people of the twenty-first century have to spiritually accept what science tells us about the world and its future. So let's extend it with something that may seem quite trite: no matter how far we peer into space, God is nowhere to be seen. We knew it well before Yuri Gagarin made his first journey into outer space on April 12, 1961. But I wonder how astronauts living in the International Space Station feel when they pray "Our Father, who art in heaven ..." while being right in the middle of the symbolic place where many people down on Earth used to situate God, physically speaking, for thousands of years.[20] In the Old Testament, God almost always manifests Himself from above (a significant exception being the Burning Bush episode). For Thomas Aquinas (1225–1274), God's dwelling was indeed the "empyrean heaven," whereas the zone between Earth and the moon was purgatory. Men and women in the Middle Ages could look up and admire His eternal resting place, and even if they weren't naïve enough to believe it, it was at least a huge help for their imagination. They could raise their eyes skyward, trustfully murmuring: "Our Father, who art in heaven ..."

In comparison, modern man has apparently nowhere to project God's presence, for science has evicted Him from His previous abodes. Neither the past nor the future, neither the infinitesimally small nor the infinitely vast, is

20 If they do pray at all, but as the reader will have already guessed, this is only a thought experiment.

a complete mystery to us—it is then at the very depths of our souls that we will have to go looking for the Invisible World. While flicking through scientific magazines, we can locate Earth on a map of the visible universe, this tiny blue round house situated in the Milky Way (200 billion stars), itself a mere hamlet within the local group of 50 galaxies, itself a small town of the Virgo cluster (1,000 galaxies), a modest suburb of the Laniakea Supercluster (100,000 galaxies), which is a supercluster among other super clusters, 500 million light-years wide, with galaxies converging along gigantic gravitational canyons toward the Great Attractor, a highly dense region irremediably pulling them in at the speed of 630 km per second. This does not necessarily make God less present or less real for those who believe in Him, but it changes the way we perceive ourselves: more lonely, isolated and defenseless as ever before. Here maybe lies the great irony of a modernity that has pretended to make us almighty while allowing us to realize how small we were.

For thousands of years, what we believed about the world and the universe was more or less stable. The geocentric model, which was probably no more elaborate than what Stonehenge people thought to know about the starry sky (and they knew a lot already), remained almost unquestioned during 2,000 years, until Copernicus started to seriously undermine it in the sixteenth century. Indeed, the religious explanations of the world changed much quicker than scientific theories, and those changes (the emergence of Christianity and Islam for instance) had nothing to do with scientific progress. For the past 500 years, the scientific revolution brought on a radically new vision of man and the universe, which many people have not yet fully absorbed, even those who consider themselves as educated Cartesians. The world as humankind has perceived it since prehistoric times has come to an end, one among the other ends I have mentioned before.

That we are troubled by the fact that our faith may waver through time is understandable but should not really surprise us: we already benefit from a highly documented historic tapestry displaying the way human beliefs have evolved, involving spirits, myths, gods or a single God. Indeed, humankind has experienced many forms of religion: animism, shamanism, polytheism, monotheism, prophets, messiahs, illuminations, mysticism, ancestral wisdom, deism, theism, theosophism, syncretism and New Age beliefs. The youngest of the world's major religions, Islam, is already nearly 1,400 years old. Man's faith has been around for such a long time that it is hard to imagine it could take on any new radical forms in the years to come, and this also participates in the end of our world. There is no burgeoning religious movement today that could hope to bring about such a change in the nature of faith as did the arrival of Christianity, which represented such an upheaval in the relationship between man and God by placing monotheism and compassion at the heart of a polytheistic, naturalistic and cruel world (well, cruelty endured). We must not forget the times when the influence of monotheism, dominant today, was anecdotal, with the exception of a few brief forays such as the cult of Aten or Ancient Judaism, built on Yahweh's triumph over other gods.

For the moment, I believe the toughest test for today's major religions will not come from their relocation to another galaxy, but rather from the simple fact that they are getting older. All of these religions have undergone an age-related crisis that is not measured by the mere impact on the number of followers. No religion is expecting any arrivals, only returns. In different ways, Christianity and Islam have suffered (and always will suffer) the ravages of time, for they are both connected to a divine intervention occurring at a specific date in history. Their respective faiths are based on the acceptance of testimonies concerning increasingly antiquated events.[21] What will the Bible's message hold in 1,000 years? From the moment that we are able to predict the death of the sun, we will no longer be able to escape our own inner torment about the meaning of our most fundamental beliefs, and this for centuries, nay millennia, to come. I am not ashamed to say that I am rattled by the idea that the very foundations of trust I held in life, and especially the afterlife, may be nothing more than a chapter in high school religious history books in the twenty-fifth century, in the same way we now consider the religious beliefs of ancient Greece and Egypt. This reminds me of the fictional dialogue between Jesus and Roman tribune Clavius in the film *Risen*: "What frightens you?" asks Christ, to whom Clavius replies: "Being wrong. Wagering eternity on it."[22]

Buddhism, Taoism and other oriental religions for their part may appear closer to the global vision presented by modern science, but this does not necessarily mean that they are any closer to the truth, if indeed there is one. Judeo-Christianity satisfies our need to personify God and the eternal sanctity of each individual, but Buddhism is quite well aligned with the cosmic vision of contemporary astrophysics. For example, concerning the theory of multiple universes (the idea that our universe is just one among many others), the Gaia theory (which states that Earth is a living organism combining life and matter)[23]

---

21 For "revealed religions," the resulting worry may be formulated as follows: has God definitively ceased to intervene in human history?

22 Directed by Kevin Reynolds, 2016.

23 Without ever validating the hypothesis of a symbiotic relationship between life and matter, studies have shown that life has influenced our planet's fate, either by modifying its atmosphere or by accelerating tectonic plate movement.

and the theory of successive universes (recurrent Big Bounces resulting from Big Crunches, *etcetera ad infinitum*). In my view, the cabalistic and mystical component of Judaism also appears to have a good chance of survival in the age of interstellar travel, having already abandoned the purely historic nature of the Torah in favor of an approach that uses the sacred text to question God and His creation. Questions will survive all wars, earthquakes and every scientific discovery.

Who will be the last God on Earth? This is an intriguing question. In any case, it is inevitable that religions will change shape and that the basis of humankind's faith will also evolve. Their survival in fact depends on this, for life is change. Anything inert will die. If we ever leave Earth, we will have to mourn our birthplace and do away with our psychological crutches and terrestrial miracles, histories and superstitions. This will not stop the astronauts who venture into the intergalactic void from taking religious objects with them, as seen in the already mentioned film *The Martian*, in which Matt Damon finds some comfort in a small wooden crucifix. But he also sacrilegiously scrapes a few shards off to start a fire and save his life, which could be a spiritual metaphor: humanity will have to say goodbye to the medieval faith of old, made of simple stories on stained-glass windows, the same faith of my childhood catechism.

As surprising as it might seem, this will be harder than overcoming our divisions. In 1600, Giordano Bruno was burned at the stake for having claimed in 1584, among other things, that there was an infinite number of earths orbiting an infinite number of suns. In that, he questioned the initial pact proposed by man to God in Genesis: tell me that I am the center of Creation and I will obey you. According to Freud, science already inflicted three "narcissistic wounds" on human self-esteem: the Copernican revolution, which put an end to the belief that Earth and humanity physically stood at the center of the universe; Darwin's theory of evolution, which makes man the natural descendant of the monkey; and Freud's own demonstration of the submission of free will to unconscious forces. The existence of other inhabited worlds would be more than a wound, a fatal blow to the singular place of man in the universe: if I am not the center of Creation any more, then why should I obey you?

Twenty-second-century people will definitely be better prepared for the prospect of visiting these places. Nevertheless, the cosmogony of the major revealed religions would not be immediately favorable to the possibility of interstellar travel. The idea that man's fate lies elsewhere than on this Earth can, in the here and now, only serve to upset the framework of most religions as they are perceived by their followers today (I admit this may not be so true in the case of more "cosmological" religions such as Buddhism, but Buddhism also teaches us to renounce to our desires, so this should include space conquest). No sacred scriptures seem to "confront the reality of interstellar travel."[24] In one way or another, leaving Earth represents a spiritual revolt, for it implies turning to the Creator and saying to him: "Lord, I know you've given us this beautiful planet but I want to leave this place. I'm not going to wait until the Apocalypse. I don't want to die there, I do not accept my fate. I want to know where I come from, where the universe ends, and why I am here. I just cannot be content with living a good life and follow your rule, and I cannot be forever grateful for something I never asked for. I am getting really bored here, I want to explore, and above all things, I want to know." In fact, leaving Earth will be like eating the forbidden fruit picked off the Tree of Knowledge, thus committing the Original Sin again and the biggest prison break in history.[25]

I don't think that resistance to research and projects about space exploration will originate from organized religious institutions, but by unconscious mental attitudes that may associate the idea of a departure from Earth with the ultimate transgression, the ultimate Promethean revolt and humankind's ultimate sin against God. Such a departure would be easier to accept on a religious level if we had to leave Earth because our existence was under threat, but this will not be the real motive for our desire to go elsewhere. Man will leave because he wants to explore, see something new and put himself to the test: all crimes against acceptance. Indeed, have we ever envisaged anything bigger than this about our human condition? In any case, I feel that religions would be wrong to fear a departure from Earth. On the contrary, it is our very confinement that runs the risk of leading humanity to nihilism and to reject a Creator who would be seen as malevolent rather than benevolent: a monster who would keep us locked up like animals in a global zoo, watching us through the bars in amusement as we kill each other like in the circuses of Ancient Rome.

We are still free to believe in any religious explanation regarding why we exist. We are still free to believe that Gods and spirits care for us in our lives. We are still free to believe that humanity was desired by God and that the whole universe was designed to welcome us. But the comforting myth and mysteries of old are gone forever. Seen from above, our situation on Earth appears very

24 As old Professor Brand (played by Michael Caine) puts it in *Interstellar*.

25 When reading the Bible, I have the feeling that this rebellion spirit is encouraged, that it is not only a right or a possibility but also a duty. After Jacob wrestled all night with a mysterious angel representing God, his name is changed to "Israel," which means "to struggle with God." Why should this struggle be limited to intellectual debating?

much like in *The Hunger Games*[26] or in *The Maze Runner*:[27] we were thrown on a wild planet with no way out or no indications about who we are, where we come from, why we're here and what awaits us in the Other world. If God exists, if He deliberately wanted us to become intelligent beings, then He is pushing us out of the nest, and we cannot expect any help from Him for that. We are free to go, but we have to earn the money for it. We can rely only on ourselves. We are on our own, and this is the exact meaning of freedom. Making use of this freedom to leave Earth will be the most audacious thing humanity could ever accomplish. But are we ready to accept the spiritual upheaval that such a move implies?

---

26 Suzanne Collins, *The Hunger Games* (New York: Scholastic, 2010).

27 James Dashner, *The Maze Runner* (New York: Delacorte, 2010). This book tells the story of a teen, Thomas, who wakes up in a glade in the center of a gigantic maze, with no memory of his past and no idea about the reason of his presence there, finding himself a resident in a community of teenagers in the same situation who are trying to find a way out. This seems to me a pretty good metaphor of the human condition.

CHAPTER 9

TRAVELING THROUGH SPACE (ALIVE)

Is it reasonable to imagine that the human race will one day leave this solar system and travel to another planet? The Alpha Centauri star system, which is the closest from our Solar System, is made up of three neighboring stars: Alpha Centauri A, Alpha Centauri B and Alpha Centauri C, also known as Proxima Centauri. On March 27, 2015, researchers announced the possible discovery of a planet situated in the habitable zone of a star close to Alpha Centauri B, at a distance of 4.37 light-years from Earth, a discovery that will need confirmation in the years to come.[1] On August 24, 2016, astronomers confirmed the discovery of a rocky planet orbiting Proxima Centauri and situated in its habitable zone, with a mass at least 1.3 times that of Earth.[2] Assuming that one of those planets could support or shelter life, can man overcome the problem of distance?

As far as it may seem, the perspective of such a travel should not be considered a foolish dream. The technology required to accomplish such a feat has yet to reach operational levels, but even in its infancy, the progress so far is huge. Back in 1909, when Louis Blériot completed the first flight across the English Channel (on a plane that looked much like a pedal boat with sails), who would have thought that only 33 years later, we would have built the first nuclear reactor capable of carrying out a sustained and controlled chain reaction? Who would have believed, during the onset of the Great Depression, that man would set foot on the moon only 40 years later? And who, back in 1941, in the heart of

1 Brice-Olivier Demory et al., "Hubble Space Telescope Search for the Transit of the Earth-Mass Exoplanet Alpha Centauri B," *Monthly Notices of the Royal Astronomical Society* 450, no. 2 (June 21, 2015), 2043–51. The formal identification of exoplanets is a slow process, given the necessity of conducting observations over several years.

2 Guillem Anglada-Escudé, P. Amado, J. Barnes, et al., "A Terrestrial Planet Candidate in a Temperate Orbit around Proxima Centauri," *Nature* 536, no. 7617 (2016), 437–40.

the Second World War, could have imagined that a Soviet lander would reach the planet Mars only 30 years later?

The National Aeronautics and Space Administration (NASA)[3] has tentatively predicted that the first manned mission to Mars will occur in the 2030s,[4] a voyage that will first require solutions for a number of problems related to manned space travel: human relationships in confined spaces,[5] protection against cosmic rays, the body's capacity to adapt to weightlessness,[6] landing difficulties and autonomous survival for more than two years, of which 500 days will be on the red planet itself. The Mars One project, which aims to colonize Mars, has already garnered 200,000 potential candidates.[7] Even if only 10 percent of these volunteers were truly ready to go on this journey of no return, the number would be extremely encouraging. Humanity is raring to go.

In June 2014, NASA revealed a futuristic presentation of a spacecraft equipped with a warp drive,[8] named the IXS-Enterprise, inspired from the famous TV series *Star Trek*, capable in theory of reaching faster-than-light speeds. Of course, this was nothing more than a communication operation aiming to stoke the fires of our imaginations, but in terms of interstellar travel, what appears impossible today may not be for much longer. Take two emblematic examples of obstacles to space exploration: propulsion and astronauts' survival in space. As far as propulsion goes, serious research has already begun on a number of

---

3 Today we all know what NASA is, but one has to think about those reading this in the thirty-third century.

4 The vessel planned for this journey is the *Orion Capsule*, which has already performed its first test flight in 2014.

5 In summer 2015, NASA placed six individuals in an 11 m^2 isolated dome in Hawai'i for one whole year in order to test living conditions similar to those on Mars (dehydrated food, walks in spacesuits, etc.).

6 The record for the longest amount of time spent in space is currently held by Russian cosmonaut Valeri Polyakov, who spent 14 months in the international space station from January 1994 to March 1995.

7 *Mars One* is a project to set up a human colony on the planet Mars starting in 2024, an idea conceived by Dutch engineer Bas Lansdorp.

8 Or the Alcubierre Warp Drive, named after Mexican physicist Miguel Alcubierre, who, in an article published in the *Journal of Classical and Quantum Gravity* in 1994, suggested that instead of accelerating, a vessel could travel by contracting space ahead of it (thus bringing its destination closer) and dilating space in its wake. This would protect passengers from suffering the effects of acceleration, which would otherwise kill them. The energy necessary for such a concept is, for the moment, not within our grasp.

theoretical topics: nuclear pulse propulsion,[9] the nuclear fusion rocket,[10] the antimatter drive,[11] solar sails,[12] the EmDrive, and so on.[13] In 2015, a company called AARC (Ad Astra Rocket Company), financed by NASA, presented the first results of an innovative thrust system, already vacuum-tested, with a name taken straight out of a sci-fi film: a *Variable Specific Impulse Magnetoplasma Rocket*, which could generate enough energy to reach Mars in 39 days. Clearly, we are not short of projects in the pipeline. There is no doubt that, in the long term, man will be able to come up with technology that today seems inconceivable. But why do we not believe this? Because the human spirit is fundamentally incapable of accepting anything new until it has been witnessed in the flesh.

Astronauts' survival over long periods in space may be extended by a process made popular in science fiction: hibernation. In October 2014, a

---

9 This was the first technology proposed in the 1950s for interstellar travel (Project Orion). As its name suggests, a spacecraft's propulsion would come from a series of successive nuclear explosions.

10 Made popular by Project Daedalus (1973), it would enable us in theory to reach 10 to 12 percent of light speed. Beyond the problem of amassing the sufficient fuel (Helium-3), which would require its own mission to pump it from Jupiter's atmosphere, the issue of mastering nuclear fusion would still remain. However, the International Thermonuclear Experimental Reactor (ITER) is no longer the only project progressing in this field. In 2015, three laboratories announced they had made some promising progress: the Lockheed Martin magnetic cylinder, the mega-laser at the Lawrence Livermore laboratory's National Ignition Facility (the United States) and Italian engineer Andrea Rossi's cold fusion reactor, E-Cat (Energy Catalyzer).

11 When a particle of matter meets an antimatter particle, they annihilate each other by giving off a quantity of energy that corresponds exactly to their mass (making a yield of 100 percent). Antimatter has already been created in laboratory conditions, in particular by Professor Jeffrey Hangst at CERN. It could in theory be the best interstellar fuel (as it generates a lot more energy than any other known fuel), but there are obstacles still to be dealt with: the considerable cost, storage, conversion of annihilated matter to dust and the production of sufficient amounts in a limited time. As well as this, there is the not-insignificant problem that more energy is required to create antimatter than it actually produces itself.

12 Solar sails use the pressure from solar or stellar radiation to accelerate (a laser can also be used). Prototypes either have already been tested or are currently in development.

13 Which stands for electromagnetic drive or electromagnetic resonant cavity thruster (*Star Trek* is now clearly old hat). According to its inventor Roger Shawyer, it is a way of providing a craft with propulsion without fuel by using microwaves trapped in a tapered cavity whose reflections provide thrust (according to a physical principle that is yet to be fully explored). Tests carried out by several laboratories have so far provided meager results.

company called Space Works (also financed by NASA) proposed a way of placing astronauts in a state of controlled hypothermia by lowering their body temperature by 3 to 5 degrees Celsius, thus slowing down their metabolism. This protocol has already been applied in medicine in order to save the brains of individuals following accidents, but the immersion of human beings into a state of hibernation for months, or even several years, would be a whole new ball game. However, nothing prevents us from imagining that we may one day succeed. A number of mammals are capable of transferring their bodies to a state of deep sleep without any assistance, and I am convinced that we will one day not only discover a way of triggering a similar state for humans but also learn considerably more about living together in closed and confined spaces. Other obstacles remain for interstellar travel: the reaction of the human body to cosmic rays and weightlessness, except where artificial gravity is generated; the procurement of water, oxygen and food; and the full recycling of everything produced up in space. However, when trains first appeared, there were concerns that humans couldn't handle such speeds and acceleration. Just a little over a hundred years ago, some of the world's most respected experts were adamant that Earth could not sustain 1 billion people. For many years it was believed the four-minute mile was a physiological impossibility. History shows that we are, as a species, extremely adept not only at breaking through such seemingly unbreakable barriers but also at traveling so far beyond them that our forefathers' beliefs on those barriers become a matter of baffled amusement.

Interstellar travel may represent a mammoth undertaking for our minds and bodies, but we have shown ourselves to be an exceptionally adaptable species, capable of settling in the most hostile of environments. In Oymyakon, Siberia, around eight hundred inhabitants endure the month of January at average temperatures of −46°C, so cold that boiling water freezes before it even hits the ground. There are 3 hours of daylight per day in December, 21 hours in June. In many African villages, the average temperature never drops below 40 degrees Celsius in summer. Some people live in countries where there is full daylight a large part of the year, whereas others survive in places where darkness reigns. Apart from temperate climes, Earth does not appear to be such a hospitable place for humans who, early on in history, had to show huge ingenuity to survive the harsh elements in places where other animals failed to adapt and did not survive. We would therefore be able to cope well with all kinds of climatic and geological configurations.

This is how I imagine the future to unfold: I believe that we will colonize space in the same way we colonized Earth. From island to island, continent to continent, adventuring along river banks, then sailing the open sea, often with no certainty about our destination. We will send men and women to Mars. We will set up a lunar base. We will learn how to exploit the resources of the solar system, which will become part of our economy and will serve to finance a large part of the conquest of space. We will explore Enceladus, Europa and other moons. We will send probes to the closest stars.[14] We will identify exoplanets of similar size to Earth, possibly sustaining life or even already hosting life-forms. Like the sky and the sea, space is in perpetual movement, shaken by solar winds, gravitational currents and natural cataclysms. Here we will discover unknown physical or cosmological phenomena that we will use to travel (space-time deformations acting like interstellar space highways?), in the same way that we learned to control the skies and navigate the seas. The first interstellar spacecraft will most surely be constructed "up" in space.

Let's have a dream. One day, maybe at the end of this century, a team of astronauts, not more than six or seven, will board this vessel. The event will be global and humanity will hold its breath. The men and women taking part in this adventure will have undergone a ruthless selection process, one of trial, error and bitter failures. The organization of their journey will have been made possible by the largest international cooperation program ever known, a project bringing together almost every domain of science and knowledge. They will be the very first Team Earth, and their launch into space will be a battle cry for humanity, a gauntlet thrown down to the astral void and to adversity, and a time of both fear and immense pride. After years of outstanding travel, these astronauts will spot another star. They will then approach a rocky planet, which may be just as blue as ours, seen in the reflection of their visors. They will cross its atmosphere and land on a hillside or by a lake, or in the middle of a forest full of unknown animals.[15] They will set up their base camp and then send a message back to Earth, saying, "We have arrived."

---

14 In April 2016, cosmologist Stephen Hawking and Russian millionaire Yuri Milner announced a plan, called Breakthrough Starshot, to send light-propelled nanocrafts toward Alpha Centauri.

15 I imagine it very much like the grandiose and heavily forested calderas of La Réunion (an island in the Indian Ocean), where you would be hardly surprised to meet a pterosaur.

EPILOGUE

I could have chosen to close this book with a happy ending involving humankind's glorious lift-off into space, after having defeated the Great Filters of all kinds. But the pages of this existential-spatial journey would not be complete without confronting one final question: so what? Will a departure from Earth enable humanity to reach our goal, if indeed we have one? Will it provide answers to our deepest questions and darkest fears? Will it bring us any closer to the truth about who we really are, why we exist, and what awaits us in the great beyond?

When I talk about this idea of traveling to another Earth, I am faced with three different types of reactions, vibrating like three electrical poles of the human psyche: the neutral, the negative and the positive. Those to whom I have spoken will, I trust, forgive me for caricaturing them a bit. Those in the neutral camp look at me with surprise, but with no real sparkle in their eye. They are indifferent to questions about the technological feasibility of such a journey, for they simply cannot see the point. They may have loved *Star Wars*, but feel nothing when watching *Interstellar*. For them, life is here and now, and the question of humankind's destiny does not extend beyond the life of their children's children. They believe that humans will disappear anyway well before they can travel to another star system. Those in the negative camp fix me with an inquisitive glare: God did not create man for him to reach the heavens through any other vehicle but prayer—we were made to cultivate the land with sweat dripping from our brows, suffer great pain in childbirth and follow His commandments. For them, we are destined for tragedy, and leaving Earth is more likely to destroy us than to save us, as if we were climbing the Tower of Babel itself. Of course, they don't put it like that, but it is from this unconscious conformist source that they draw their instinctive disapproval, even though some may be ardent atheists (the legitimation of any existing order being to my view a psychological trait that is not necessarily rooted in religious dogma). This leaves us with the positive camp, for whom the mere words "interstellar travel" excite and thrill, conjuring imaginary scenes of spaceships flying off to conquer far-off galaxies. For these science-fiction fanatics, the only way for humankind to be existentially fulfilled is through quests, exploration and technological progress. Everything else, in both the abstract and concrete sense, is nothing but literature.

Wherein lies the truth? Probably somewhere in the middle, as always. One unseen risk faced by hardened adventurers is that of being stripped of any desire to ask metaphysical questions. Their physical struggles may cause them to lose themselves in the bittersweet comfort of believing that only the biological continuity of our species is at stake. For me, their thirst for space adventures seems to ignore the spiritual dimension of the fate of humanity. Conversely, we are not pure spiritual actors, bound to play a tragic role on the terrestrial stage. We're also natural beings born to an Earth that has shaped our bodies and a large part of our souls. We were not just parachuted in one fine morning among the forests and beasts of a pristine Earth like in Genesis. The way we perceive our condition and envisage our future is first and foremost inspired by the extraordinary sensory machine that governs our thoughts, the result of millennia of evolution, without which we wouldn't have the slightest notion of the world that surrounds us. Our fate is closely knit to that of our ecosystem. Yet we're not easily prone to perceive ourselves as part of nature, for in the Western world, we have inherited Judeo-Christian thought in which the cosmos and nature are practically absent or radically separated from man, and where the only thing that counts is our relationship with God and thy neighbor. We may be different though from the rest of creation, but not exceptional to the extent that our environment has no bearing on our spiritual evolution. We have been shaped and modeled by our experience on Earth, in the same way that a soldier comes home completely transformed by war. Now that we can locate ourselves in the evolution process and on the space map, we should pay greater attention to our destiny as a species and not only as mere characters of a metaphysical scenario.

Let's go further in the future. When the day comes when space is conquered, when the discovery of new planets and galaxies will generate no interest any more, what will be left for humankind to accomplish? Will we have reached such a level of controlling matter and space that we will think we have risen to the status of Gods? There will probably always be a limit to our full understanding of why we exist in the first place, as there will always be a veil between us and the ultimate secrets of nature. This is why both religions and ancient philosophies have not had their final word. Antique cosmogonies, classical Greek thought and the narrative in Genesis still create such fascination for us because we can find in them reflections and ideas that were not yet reduced to mere results observed under the scrutiny of telescopes and microscopes. Today, our capacity for thought is biased, influenced by the mass of knowledge accumulated over the centuries—we hopefully no longer believe that storms come from the wrath of God, but our primal curiosity regarding the mysteries of the universe has certainly been stifled. For all our machines and expertise, the path that lies ahead is just as obscure as it was 2,000 years ago. Is there anyone today who can perceive the world in the same way as a philosopher 400 years before Christ? We have so many answers now that I sometimes feel we are no longer capable of asking the right questions, especially as regards what is good for us, fascinated as we are by the exploits of science.

Our starting point is death, and therefore life. We may go far as a species, but we will always be limited by our individual life spans—even though future generations will be able to extend it (and I would find it hard to use the word *life* when describing the transhumanist project to digitalize the human spirit). The conquest of space will surely push us beyond all our physical, intellectual, existential and spiritual boundaries. But no matter how far we flee into space, we will always remain prisoners of time. The first man (or woman) to die on another planet will be no more comforted as he (or she) dies than we are when our own time comes. He (or she) too will look skyward, perhaps at a different-colored sky with more suns or moons, and will be no less fearful than us at the thought of losing, in a flash of light, the one thing we believe is truly ours: our life, our essence, our friends and family, our significance in the lives of others and maybe even our soul.

Whatever we discover on our future voyages, neither we, our children nor our descendants will be in any way better equipped to deal with our final journey across the river Styx. Death is the human condition's most unwanted visitor, one which all our beliefs attempt to explain away, to ignore or even to cheat. But the journey into space will have a considerable impact on the understanding and transformation of the human condition. It will also be an inner quest and a spiritual quest, much like any voyage, taking us back to the days when man first decided to leave the African continent to explore the world.[1] Images provide us with an idea of the immensity of space, but we are yet to genuinely experience this environment. Regardless of how well they are trained, those who will have to watch from their spaceship as Earth disappears from view behind them will surely be anguished at such a sight, and humanity at large

1 As Stanley Kubrick admirably depicted it in *2001: A Space Odyssey* (1968), to which *Interstellar* so brilliantly echoes.

will be just as overwhelmed. They will see the Absence, and even if they take a small religious object or a holy book with them on board, they will have to find the light within themselves, for there will be absolutely nothing in the icy void out there to provide them with any comfort whatsoever. The account of this incredible journey will surely become the bedside book of the new Earth's schoolchildren.

If the colonization of another Earth only triggers further questions instead of bringing us answers to the whys and hows of our existence, will it offer up any real hope for humanity, beyond that of our species' survival? Well, yes, for beyond the fact that it will be our greatest adventure since we decided to stand on two legs, the journey into space will never materialize without considerable scientific breakthroughs, serving to extend our knowledge about the universe. And science, whatever may be said, pushes us to meditate on issues that are beyond its purview and that we would in vain try to remove from its influence. For instance, the power of the biblical version of Creation wanes in the face of the scientific version of the Big Bang, precisely because its resemblance (albeit tenuous) leads to its inevitable substitution. We may be proud to know that the world was not created in six days, but it was just so much more convenient when that moment was still stewing in the mystery of Genesis. The good old days when science took care of the hows and religion of the whys are over. The progress in quantum physics and cosmology has turned everything upside down: science can no longer dodge questions such as "reality" or what was existing before the Big Bang, which were not long ago considered as only metaphysical questions, whereas religion cannot ignore the findings of biblical criticism and archeology. To be able to scientifically pinpoint the origin of the universe leads science to question the origin of that origin. When we examine the issue of the beginning of everything, the question of "how?" very quickly becomes "what?" or "who?" Who set the parameters to launch the creation software? Who decided on the proportion of elements of which the universe is composed? Who coded the world, or life? Eternity, infinity and nothingness all take form in our particle accelerators, microscopes and telescopes, madly waving at us, then guiding us like spiritual wreckers toward a gaping and bottomless abyss of agonizing questions.

The paradox of quantum physics (but also of mathematics and astrophysics) is that is has reenchanted[2] classical mechanistic science by extending our knowledge of the unknown. While actively searching for more certainty about

---

2 To use the term coined by Ilya Prigogine.

the nature of things, scientists eventually encountered more uncertainty. In other words, we now better know what we cannot know. Heisenberg's principle of uncertainty has taught us that we cannot simultaneously identify both the speed and position of a given particle, thus putting a stop to the deterministic theories that had excited the scientific community since the eighteenth century. Gödel's incompleteness theorem poses that any axiomatic system must contain propositions that can be neither proven nor refuted, establishing a radical limit on the mathematical pretention of building a completely self-justifying system. In cosmology, the Planck wall makes it impossible to conceptualize the way the universe functioned before an age estimated at 10^{-44} seconds, the barrier before which the physical variables reach such extremes that they can no longer be described by any laws known to physics. Measuring the age of the universe has at the same time led to the calculation of its theoretical visible limits as 47 billion light-years (the light from far-off stars has still not had the time to reach us—this number takes into account the theoretical expansion of the universe). For the moment, we have no idea as to how we will ever know what makes up dark matter or dark energy, whether the universe is finite or infinite, how it will end or what came before the Big Bang. Science is constantly shaking up the science that went before, and our certainties of today will be overturned tomorrow, giving birth to fresh uncertainties.[3] This is the essence of our human condition, and the very purpose of faith: we will not, by ourselves, pass through the looking glass.

The astronauts who will head off into the astral void will be like hermits in the wilderness, and only the strongest mystics will survive this journey, for they will have to face the primordial nothingness of the human condition all on their own. They will take with them little more than we will take to our own graves. When they leave Earth, we will all say a symbolic farewell to the places we have cherished and the people we have loved. A part of us will die with those astronauts' departure,[4] without losing hope in another life, in another galaxy far, far away, another dimension, beyond space and time. The only wisdom that

3 Such is the case of Heisenberg's principle of uncertainty, whose influence is already being doubted since the work of Japanese physicist Masanao Ozama in 2003, or Einstein's theory of relativity, which will be tested through several major experiments to be conducted in the years to come, such as the Pharao cold atom space clock.

4 The interstellar voyage as an allegory for a departure toward the darkness of death was beautifully captured in the film *Interstellar*, including a Dylan Thomas poem written for his dying father, published in 1951, the first lines of which are quoted by Professor Brand at the moment the spaceship *Endurance* takes off: "Do not go gentle into that good night / Old age should burn and rave at close of day / Rage, rage against the dying of the light."

has been all-pervading since the dawn of time is that life is only ever real in the present moment and that we must cherish each of these instants and every person who fills these moments with joy as an immeasurable treasure that we will never fully possess. This, dear reader, may well be the only wisdom that truly deserves to be retained from this book.

> With the drawing of this Love and the voice of this Calling
> We shall not cease from exploration
> And the end of all our exploring
> Will be to arrive where we started
> And know the place for the first time.
> Through the unknown, unremembered gate
> When the last of earth left to discover
> Is that which was the beginning;
> At the source of the longest river
> The voice of the hidden waterfall
> And the children in the apple-tree
>
> Not known, because not looked for
> But heard, half-heard, in the stillness
> Between two waves of the sea.
> Quick now, here, now, always—
> A condition of complete simplicity
> (Costing not less than everything)
> And all shall be well and
> All manner of thing shall be well
> When the tongues of flames are in-folded
> Into the crowned knot of fire
> And the fire and the rose are one.
>
> (T. S. Eliot, "Little Gidding")

ACKNOWLEDGMENTS

My gratitude goes first to my editors of the French original version, *Liber* (Québec, Canada), and especially to its founder Giovanni Calabrese, its director Micheline Gauthier and its delegate in France Sylvain Neault, who enthusiastically supported its diffusion. Publishing idea-driven books requires humility, courage, selflessness and passion, all qualities that guide the choices of this remarkable publishing house and its team. May they be warmly thanked for giving my thoughts a chance to find their audience.

I am extremely grateful to Anthem Press for allowing my book to reach the anglophone readership. It was a pleasure to collaborate with my editor Megan Greiving, whose dedication throughout the editorial process was highly appreciable. May the whole Anthem team be thanked for their remarkable involvement.

I strongly thank Mr. James Christie, who has done a great translating job. His thoroughness and concern for choosing the accurate words and phrases decisively contributed to the quality of the English version.

Author and psychologist Richard Boston, who took the time for a complete review of the first English version, provided me with very thoughtful remarks, some of which are included in the book. I thank him for his time and his frankness.

It is well-known that cats are muses for writers, and mine were no exception. More confidential is the feline contribution to space conquest. I have a special thought for Félicette, the first cat to have been launched into space, on October 18, 1963, by France. She is the only cat to have survived spaceflight, just to be sadly killed a few months later to examine her brain. Along with the sacrifice of astronauts who died on duty, let's not forget the animals whose death also served higher purposes.

Finally, I wish to thank my wife and my three children for their constant support and encouragements. The penultimate sentence of this book is dedicated to them.

Lightning Source UK Ltd.
Milton Keynes UK
UKHW011848301021
393067UK00002B/53